Machining of Hard Materials

J. Paulo Davim

Machining of Hard Materials

Springer

J. Paulo Davim, PhD
Department of Mechanical Engineering
University of Aveiro
Campus Universitário de Santiago
3810-193 Aveiro
Portugal
pdavim@ua.pt

ISBN 978-1-84996-449-4 e-ISBN 978-1-84996-450-0
DOI 10.1007/978-1-84996-450-0
Springer London Dordrecht Heidelberg New York

British Library Cataloguing in Publication Data
A catalogue record for this book is available from the British Library

Cover design: eStudioCalamar, Girona/Berlin

Printed on acid-free paper

Springer is part of Springer Science+Business Media (www.springer.com)

Preface

Hard machining is a recent technology that can be defined as the machining operation of a workpiece that has a hardness value typically in the 45–70 HRC range, using directly tools with geometrically defined cutting edges. This operation always presents the challenge of selecting a cutting-tool insert that facilitates tool life and high-precision machining of the component. Hard machining presents several advantages when compared with the traditional methodology based on finish-grinding operations after heat treatment of workpieces. Also this technology presents a great contribution to sustainable manufacturing.

The hard-materials group comprises hardened steels, high-speed steels, heat-treatable steels, tool steels, bearing steels and chilled/white cast irons. Also, Inconel, Hastelloy, cobalt alloys for biomedical applications and other special materials are classified as hard materials. These materials are in constant use by the automotive industry for bearing production and for machining of moulds and dies as well as other components for advanced industries.

Chapter 1 of the book provides the definitions and industrial applications of machining of hard materials. Chapter 2 is dedicated to advanced cutting tools. Chapter 3 describes the mechanics of cutting and chip formation. Chapter 4 contains information on surface integrity. Chapter 5 is dedicated to finite-element modelling and simulation. Finally, Chapter 6 is dedicated to computational methods and optimization.

This book can be used as a textbook for final undergraduate engineering courses or as a topic on manufacturing at the postgraduate level. It can also serve as a useful reference for academics, manufacturing researchers, manufacturing, materials and mechanical engineers, and professionals in machining and related industries. The scientific interest of this book is evident for many important centres of research, and laboratories and universities throughout the world. Therefore, it is hoped this book will inspire and enthuse other researchers in this field of machining science and technology.

I am grateful to Springer for this opportunity and for their enthusiastic and professional support. Finally, I would like to thank all the Chapter authors for their contributions to this work.

University of Aveiro, Portugal, April 2010 *J. Paulo Davim*

Contributors

Prof. Alexandre M. Abrão
Department of Mechanical Engineering
Universidade Federal de Minas Gerais
Av. Antônio Carlos,
6627, Pampulha
31270901, Belo Horizonte MG
Brazil
e-mail: abrao@ufmg.br

Prof. P.J. Arrazola
Faculty of Engineering
Mondragon University
Loramendi 4
20.500-Mondragón
Spain
e-mail: pjarrazola@eps.mondragon.edu

Prof. Viktor P. Astakhov
Department of Mechanical Engineering
Michigan State University
2453 Engineering Building East Lansing
MI 48824-1226
USA
e-mail: astvik@gmail.com

Dr. I. Azkona
Metal Estalki, Derio, Bizkaia
Spain
e-mail: iazkona@metalestalki.com

Prof. J. Paulo Davim
Department of Mechanical Engineering
University of Aveiro
Campus Santiago
3810-193 Aveiro
Portugal
e-mail: pdavim@ua.pt

Prof. Wit Grzesik
Faculty of Mechanical Engineering
Department of Manufacturing Engineering and Production Automation
Opole University of Technology
P.O. Box 321
45-271 Opole
Poland
e-mail: w.grzesik@po.opole.pl

Prof. A. Lamikiz
Department of Mechanical Engineering
University of the Basque Country
Faculty of Engineering of Bilbao
c/Alameda de Urquijo s/n
E-48013 Bilbao
Spain
e-mail: aitzol.lamikiz@ehu.es

Dr. J. Fernández de Larrinoa
Metal Estalki, Derio, Bizkaia
Spain
e-mail: jflarrinoa@metalestalki.com

Prof. L. Norberto López de Lacalle
Department of Mechanical Engineering
University of the Basque Country
Faculty of Engineering of Bilbao
c/Alameda de Urquijo s/n
E-48013 Bilbao
Spain
e-mail: norberto.lzlacalle@ehu.es

Prof. Ramón Quiza
Department of Mechanical Engineering
University of Matanzas
Autopista a Varadero, km 3½
Matanzas 44740
Cuba
e-mail: quiza@umcc.cu

Prof. José Luiz Silva Ribeiro
Department of Mechatronic Engineering
Pontifical Catholic University of Minas Gerais
Av. Dom José Gaspar
500, Coração Eucarístico
30535-610, Belo Horizonte MG
Brazil
e-mail: ribeirojls@yahoo.com.br

Contents

Chapter 1
Machining of Hard Materials – Definitions and Industrial Applications

V.P. Astakhov

1.1 Introduction: Definition of Hard Machining, Advantages and Limitations

In its broad definition, hard machining is machining of parts with a hardness of above 45 HRC, although most frequently the process concerns hardnesses of 58 to 68 HRC. The workpiece materials involved include various hardened alloy steels, tool steels, case-hardened steels, superalloys, nitrided irons and hard-chrome-coated steels, and heat-treated powder metallurgical parts. It is mainly a finishing or semi-finishing process where high dimensional, form, and surface finish accuracy have to be achieved [1].

Since its broader introduction in the mid 1980s in the form of hard turning, hard machining has evolved considerably in various machining operations as milling, boring, broaching, hobbling, and others. Developments of suitable rigid machine tools, superhard cutting-tool materials and special tool (toolholders) designs, and complete set-ups has made the metal cutting of hardened parts easily accessible for any machine shop.

The conventional solution to finishing hardened steel parts has been grinding, but there are a number of clear benefits to the machining of hard parts with a cutting tool. These have justified many existing applications that are growing in number, especially involving turning, boring, and milling [1]. Hard turning was early recognized and pioneered by the automotive industry as a means of improving the manufacturing of transmission components. Gear-wheel bearing surfaces are a typical example of early applications converted from grinding to hard machining using inserts in polycrystalline cubic boron nitride (PCBN). Case-hardened steel

V.P. Astakhov
Department of Mechanical Engineering, Michigan State University,
2453 Engineering Building East Lansing, MI 48824-1226, USA
e-mail: astvik@gmail.com

components are typical, often having a hardness-depth of just over 1 mm, giving it a wear resistant case and a tough core. Components that make use of this combination of material properties include gears, axles, arbors, camshafts, cardan joints, driving pinions, and link components for transportation and energy products, as well as many applications in general mechanical engineering [1].

Today hardened components are machined widely across many different industries. In modern manufacturing, hard machining is no more seen as the alternative to all grinding operations, although many research papers by university researchers still maintain this outdated notion. In real-world manufacturing, there are a number of applications where these two processes complement each other, thus modern machines for hard machining are often equipped with grinding spindles. Figure 1.1 shows a typical automotive part, a supporting shaft assembly (six-speed rear-wheel-drive automatic transmission). After the hardened (48–50 HRC) supporting shaft is pressed into the hard (42 HRC) cast iron flange, machining operations are used to assure perpendicularity of the flange face and shaft shoulders. As such, the shaft shoulder is subjected to PCBN turning, while the flange face is turned with a ceramic-insert tool and finished with a grinding wheel (diamond plated) to assure its flatness and surface finish. All these operations are performed on the same EMAG VSC machine.

The following benefits of hard machining (well-described in many literature sources and companies' promotion materials) have been experienced by users of the process:

- easy to adapt to complex part contours;
- quick change-overs between component types;
- several operations performed in one set-up;
- high metal removal rate;
- same computer numerical control (CNC) lathe as used for soft turning is possible;
- low machine tool investment;

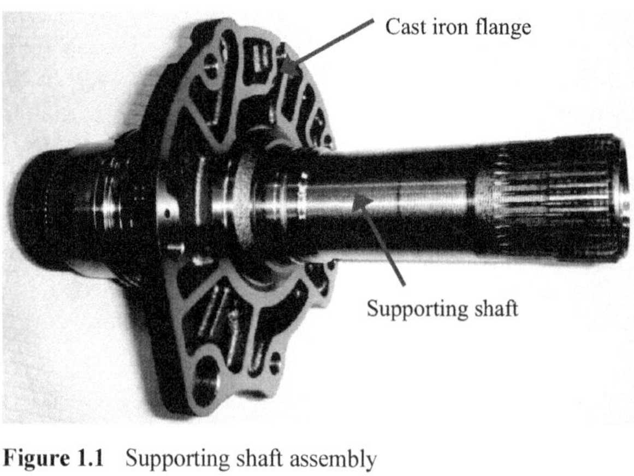

Figure 1.1 Supporting shaft assembly

- environmentally friendly metal chips;
- elimination of coolants in most cases;
- small tool inventory.

The limitation and drawbacks of hard machining are not normally listed in promotional materials and in research papers, although they should be clearly understood by end users:

- The tooling cost per unit is significantly higher in hard machining compared to grinding.
- In some cases, a part's size or geometry simply does not lend itself to hard turning. Parts that are best suited for hard turning have a small length-to-diameter (L/D) ratio. In general, an L/D ratio for unsupported workpieces should be no more than 4:1. Despite tailstock support for long, thin parts, high cutting pressures would likely induce chatter.
- In many cases special rigid machines are required for successful hard machining. The degree of machine rigidity dictates the degree of hard- turning accuracy. As required part tolerances become tighter and surface finishes get finer, machine rigidity becomes more of an issue. Machining systems should integrate a number of features to increase rigidity and damping characteristics for hard-machining applications. These include machine bases with polymer composite reinforcement, direct-seating colleted spindles that locate the spindle bearing close to the workpiece, and hydrostatic ways. Maximizing system rigidity means minimizing all overhangs, tool extensions, and part extensions, as well as eliminating shims and spacers. The goal is to keep everything as close to the turret as possible.
- The biggest question with respect to coolant is whether or not to use it in hard machining. For parts such as gears, that have interrupted cuts, it is best to machine dry. That is because the thermal shock the insert would experience exiting and entering cuts would likely cause breakage. For continuous cuts, the high tool tip temperatures that occur in dry turning serve to anneal (soften) the pre-cut area, which lowers the hardness value and makes the material easier to cut. This phenomenon is why it is beneficial to increase the speeds when cutting dry. Cutting without coolant provides obvious cost benefits as well. However, part thermal distortion, handling, and in-process gaging may present significant problems. The latter issues force the use of coolant in some applications. If a coolant is used, it must be water-based. Near-dry machining has proved to be beneficial in hard machining.
- Surface finish of machined parts deteriorates with tool wear even within the limit of tool life. Figure 1.2 shows an example.
- The so-called "white layer" formation in hard machining [2–4], invisible to the naked eye, is a very thin shell of material that is harder than the underlying material. The thickness of a white layer formed during hard machining increases with tool wear, as shown in Figure 1.3. It is most commonly formed on bearing steels and is most problematic for parts like bearing races that receive high contact stresses. Over time, the white layer can delaminate and lead to bearing fail-

ure. Most bearing manufacturers have in-house testing resources to deal with this issue. For a shop just getting into hard turning, it is recommended that random tests be performed during the first few weeks of production to determine how many parts can be cut per insert without white layer formation. A metallurgical company can perform these tests. Even though an insert may be able to cut 400 parts within specification, it may be that it has dulled to the point where it causes a white layer after only 300 parts.

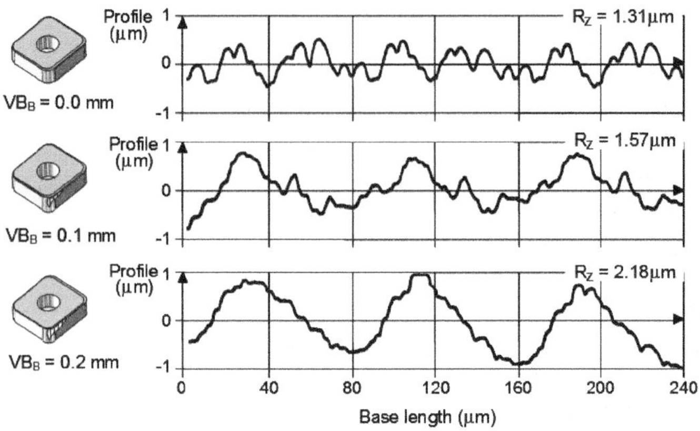

Figure 1.2 Surface finish deteriorates with tool wear. Turning AISI steel 52100 (65.5 HRC), with a PCBN tool (insert DNGA 150612, TiC binder), cutting speed $v = 140$ m/min, feed $f = 0.08$ mm/rev, depth of cut $d_w = 0.15$ mm, tool nose radius $r_n = 1.2$ mm, dry cutting

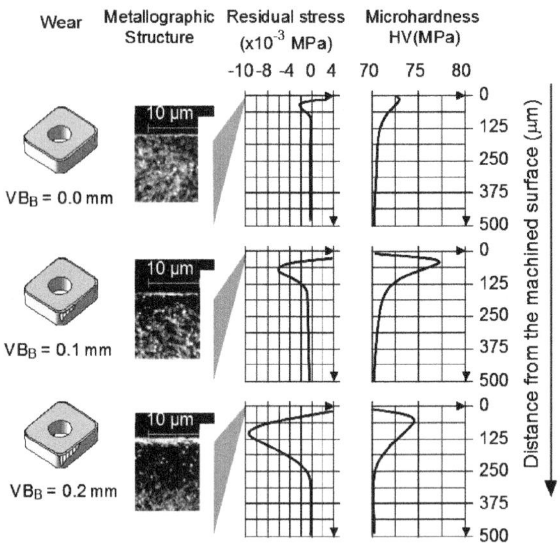

Figure 1.3 Deterioration of surface properties with tool wear. Turning AISI steel 52100 (61.5 HRC), with a PCBN tool (insert DNGA 150612, TiC binder), cutting speed $v = 140$ m/min, feed $f = 0.08$ mm/rev, depth of cut $d_w = 0.15$ mm, tool nose radius $r_n = 1.2$ mm, dry cutting

1.2 Short Critical Analysis of the Research on Hard Machining

Although a great body of literature on hard machining is available, only very few publications deal with the physical and mechanical essence of this type of machining, attempting to explain the facts known from practice [5–10]. In the author's opinion, even these few papers tried to "converge" conventional and hard machining, ignoring some obvious facts known from practice. As a result, leading tool companies have considerably different ideas about the essence of hard machining as described in these papers. Unfortunately, many researchers in the field of hard machining do not "hear" this opinion.

Nakayama *et al.* published probably the first comprehensive paper where the attempt was made to generalize the experience on hard machining [5]. They stated that "the machining of hard materials is very different from conventional machining. Many of our knowledge and theoretical works on the conventional machining cannot be applied. For the progress of machining technology for hard metals, the machining characteristics of such materials must be examined basically". Unfortunately, many later researchers did not follow this great advice.

The following summarizes the results obtained by Nakayama *et al.* [5]:

- In hard machining, the so-called saw-toothed chip is formed due to *fracture* of the work material as shown in Figure 1.4. As such, the *crack* initiates at the workpiece free surface when the work material attains the limiting shear strain. Therefore, *fracture governs the chip formation process*.
- Although the "segmental chip" formed in difficult-to-machine materials due to adiabatic shear has a similar cross-section to the saw-toothed chip formed in hard machining, *these two chip types are not the same* as they are formed by different mechanisms.
- The shear angle is very small compared to the traditional machining. It significantly increases with the hardness of the work material and weakly depends on the tool rake angle.

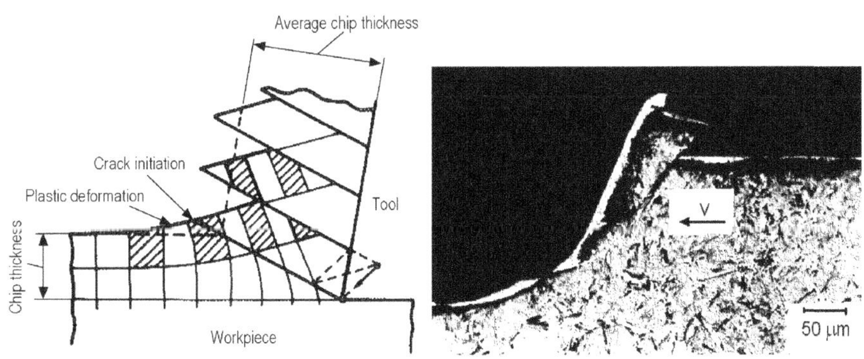

(a) (b)

Figure 1.4 (a) Model of chip formation suggested by Nakayama *et al.*, and (b) its experimental verification by König *et al.* [11]

- In hard machining the radial (thrust) component of the cutting force is greater than the tangential (power) component and the difference increases significantly with flank wear. This is attributed to springback of the work material. Large thrust force is the prime cause for size error in hard machining.
- The tangential (power) and radial (thrust) components of the cutting force depend on the tool rake angle. As such, when rake angle is zero, these components do not increase with the hardness of the work material, while when this angle is –20°, these components significantly reduce with this hardness.
- The tangential (power) and radial (thrust) components of the cutting force depend on the flank wear differently. *The radial component increases four-fold when flank wear increases from zero to 0.2 mm.*
- *The chip compression ratio* calculated as the ratio of the average chip thickness to the uncut (undeformed) chip thickness and *is approximately equal to two.*

What was not explained (discussed) by Nakayama *et al.* [5]:

- *The shape of the chip.* It is not clear why according to the model shown in Figure 1.4 the chip forms a continuous ribbon consisting of elements tightly connected to each other, while in machining of cast iron the chip forms as separate elements, although ductility of gray cast iron is normally greater than that of hardened steel. For example, one of the most widely used in hard-machining applications, and thus testing, AISI 52100 bearing steel having hardness 60–67 HRC (analogous to that used by Nakayama *et al.* in their tests) has elongation at fracture of 5 %, while gray cast iron has elongation more than 7 %.
- *The chip formation model.* It is not clear why the limiting shear strain occurs at the workpiece free surface, thus *a crack starts in this region* as it does not follow from the force model analyzed in the paper. Moreover, according to the model, the chip formation is highly cyclical, thus the cutting force and its components should vary significantly within each cycle of chip formation. Besides, the energy associated with the formation of new surfaces due to crack propagation [12, 13] is not accounted for in the model and its analysis.
- *Relatively low tangential (power) and radial (thrust) components* of the cutting force and their reduction with the hardness of the workpiece while cutting with negative rake angles, which seems to be opposite to the common perception.
- *The existence of the shear angle in machining* of a highly brittle material and its correlation with the direction of crack propagation. The visible plastic deformation of the work materials and high chip compression ratio that corresponds to 200 % of plastic deformation of the work material.
- *The source of high temperature* in hard machining and its influence on the cutting process and its outcomes are not analyzed.
- *The source of dynamic instability* of hard machining is not analyzed, although practically any research and/or technical paper as well as promotional materials on hard machining point out this issue to be of prime concern in this process.

The major ideas and model of chip formation proposed by Nakayama *et al.* [5] were successfully developed by further researchers. For example, Shaw and Vyas [6] added some mechanics to the model of chip formation showing conditions of the transformation of this model when less brittle materials are cut. Studying chip formation, Davies *et al.* concluded [9] that "segmented chip formation is highly nonlinear dynamic process that can affect cutting forces, machine deflection and surface finish". Probably the most realistic model was developed by Elbestawi *et al.* [6] where fracture and heavy plastic deformation of the chip are combined to explain chip morphology.

A CIRP keynote paper by Tönshoff *et al.* [8] attempted to summarize the knowledge gained over more than a 25-year history of hard machining. Although it presents a great detailed analysis of various aspects of hard machining and the most significant work in the field, it suffers some methodological and factual flaws that affected the thinking and perception of hard machining of many subsequent researchers. Among them, the following is of prime concern.

Although subsection 3.1 states that "The stress-strain curve of hardened steel, gained in tension test, is almost linear until fracture. There is practically no plastic deformation", section 3 "Mechanism of chip removal during hard cutting" begins with the following statement: "Applying hard cutting as a finishing process requires the generation of machined surface by *pure plastic deformation*". Even though the works by Nakayama *et al.* [5], Elbestawi *et al.* [6], and Shaw and Vyas [6] are referred to in this section and crack formation is mentioned, "pure plastic deformation" rather than fracture as the mechanism of chip formation governed the mindset of the authors of the paper. As a result, no role of fracture in hard machining is even mentioned, nor is the energy associated with formation of new surfaces [13–15] considered.

The paper attempted to analyze two basic approaches (wrongly referred to as "theories" in the paper): (a) thermodynamic theory and (b) hydrostatic theory. According to the thermodynamic theory, self-induced heating causes plasticization of the work material, which is the essence of hard machining, while the second approach relies on heavy plastic deformation and adiabatic shear as in machining of difficult-to-machine materials of high spasticity. The second approach was accepted in the paper ignoring direct warnings by Nakayama *et al.* [5] and Shaw and Vyas [6] that this is not the case in hard machining. The authors of the paper failed to explain how a hard work material having only 4% strain at fracture can exhibit more than 300% plastic deformation in machining when temperature does not affect its properties.

The first approach was denied, although it is supported by the vast majority of researchers in the field and by practical experience. The reasons for denial are based on the results obtained in a master thesis where no influence of cutting speed and thermoconductivity of the work material on force components was found. In reality, however, this is not true. When highly dynamic force components in hard machining are properly measured by experience researchers,

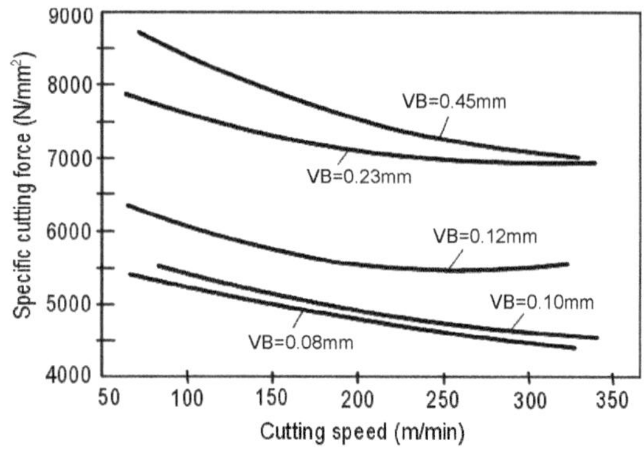

Figure 1.5 Evolution of the specific cutting force at different values of flank wear. Work materials – alloyed steel 52 HRC 1900 MPa, tool material – CBN, depth of cut – 0.25 mm, feed – 0.08 mm/rev

the cutting speed has very strong influence on these components in hard machining with PCBN, ceramic, and cermet tool materials [16–18], which is the essence of hard machining. For example, Remadna and Rigal [18] pointed out that "The specific cutting force K_{sc} decreases regularly when the cutting speed increases from 50 to 350 m/min and regardless of the wear VB", as shown in Figure 1.5.

Tönshoff *et al.* [8] provide an incorrect explanation of the fact that the radial (thrust) component of the cutting force (referred to as the passive force in the paper) is the largest component. According to them, this is because chip formation takes place mainly in the region of the corner radius and the chamfer of the cutting tool. In cutting with a round insert or with a wiper insert having a large corner radius with the wiper part, the tool cutting-edge angle is also small and the chip formation region is restricted by the tool nose radius. However, it does not make the radial component of the cutting force larger than the tangential component [19].

1.3 Factors Distinguishing Hard Machining

As pointed out by Nakayama *et al.* [5]: "For the progress of machining technology for hard metals, the machining characteristics of such materials must be examined basically". Although the body of this book examines such characteristic in great details, some distinguishing features to be addressed in the analysis of hard machining should be pointed out to set priorities in the future studies and applications of such a process. To do this, the difference in energy balance in the conventional and hard metal cutting processes should be analyzed. Astakhov and Xiao proposed the following model for energy balance in metal cutting [20, 21]:

$$P_c = F_c v = P_{pd} + P_{fR} + P_{jF} + P_{ch} \tag{1.1}$$

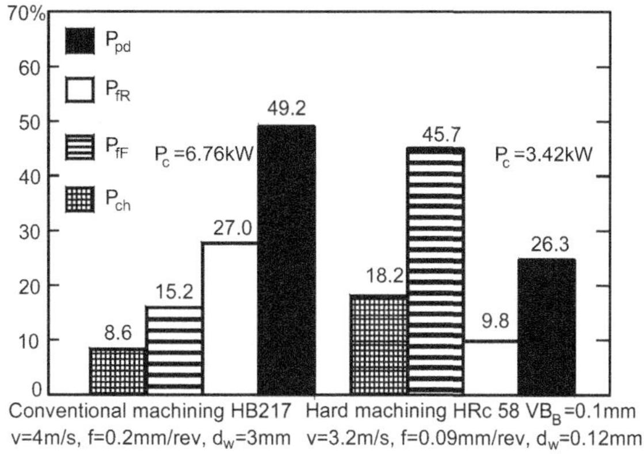

Figure 1.6 Comparison of energy balances in the conventional and hard turning of AISI steel 52100

where F_c (often referred to as F_z in metal-cutting publications) is the power (tangential) component of the total cutting force, v is the cutting speed, P_{pd} is the power spent on the plastic deformation of the layer being removed, P_{fr} is the power spent on the tool–chip interface, P_{fF} is the power spent on the tool–workpiece interface, and P_{ch} is the power spent in the formation of new surfaces.

Figure 1.6 shows the differences in energy balances in conventional and hard machining of AISI steel 52100. The following conclusions follow from this comparison:

- As seen, the power spent on the tool–workpiece interface is the greatest, which explains why the axial (thrust) component of the cutting force is greater than the tangential (power) component in hard machining, while the opposite is true in conventional machining.
- Another distinguishing feature of hard machining is significant power spent in the formation of new surfaces [13] which are never considered as a factor, although crack formation and propagation considerations are often included in the model of hard machining.
- Surprisingly in the energy balances shown in Figure 1.6, the power spent on the plastic deformation of the layer being removed is still significant in hard machining.

Understanding the physical background of these conclusions helps to reveal distinguishing features of hard machining, which are: (a) cutting force reduction with the cutting speed, (b) great axial (thrust) components of the cutting force, (c) great power spent in the formation of new surfaces, and (d) a need for rigid machining systems.

1.3.1 Cutting Force Reduction with the Cutting Speed

To understand the thermal energy (heat) influence in hard machining, a similarity number, called the Péclet criterion is useful [13, 22–25]. In metal cutting, the Péclet criterion is represented in terms of machining process parameters as follows [26, 27]:

$$Pe = \frac{vt_1}{w_w} \tag{1.2}$$

where v is the velocity of a moving heat source (the cutting speed) (m/s), and w_w is the thermal diffusivity of the work material (m^2/s),

$$w_w = \frac{k_w}{(c_p\rho)_w} \tag{1.3}$$

where k_w is the thermoconductivity of the work material (J/(m s °C)) and $(c_p\rho)_w$ is the volume specific heat of work material ($J/(m^3\ °C)$).

The Péclet number is a similarity number, which characterizes the relative influence of the cutting regime (vt_1) with respect to the thermal properties of the workpiece material (w_w). If $Pe > 10$ then the heat source (the cutting tool) moves over the workpiece faster than the velocity of thermal wave propagation in the work material so the thermal energy generated in cutting due to the plastic deformation of the work material and due to friction at the tool–chip interface does not affect the work material ahead of the tool [28]. If $Pe < 10$ then the thermal energy due to the plastic deformation and due to friction makes its strong contribution to the process of plastic deformation during cutting as it affects the mechanical properties of the work material.

As an example, consider machining of AISI 1040 steel under the typical machining conditions: operation – turning; tool – MTJNR-1616H-09 (ISO 5608:1995) with a carbide insert; cutting speed $v = 3$ m/s (180 m/min); cutting feed $f = 0.25$ mm/rev. Thermal diffusivity of the work material is $6.67 \times 10^{-6}\ m^2/s$. For the J-style tool holder, the tool cutting-edge angle is $\kappa_r = 93°$, thus the uncut chip thickness calculates as [29] $t_1 = f\cos(\kappa_r - 90° = 0.25\cos(93° - 90°) = 0.24965 \approx 0.25$ mm. Thus, the Péclet criterion calculates as $Pe = (3 \times 0.25 \times 10^{-3})/6.67 \times 10^{-6} = 112$. As $Pe >> 10$, one should conclude that the thermal energy generated in cutting due to the plastic deformation of the work material and due to friction at the tool–chip interface does not affect the work material ahead of the tool. In other words, the cutting process in this considered case is a cold-working process. Thus, there is no need to use temperature-dependent work material models (for example, the Johnson–Cook model [30]) in the numerical modeling of the considered process.

However, this is true only for the pure orthogonal cutting, where the tool never passes the same, or even the neighboring point of the workpiece more than once. In practical machining operations (turning, milling, drilling, *etc.*), the feed is used to generate the machined surface. As such, the cutting tool advances into the workpiece with the feed velocity, which is considerably smaller than the cutting velocity so that the residual heat from the previous pass might significantly affect the cutting process on the current pass. This might happen, however, if the residual temperature is high. Silin showed [31] that in conventional machining, the temperature rise in the current pass due to residual heat from the previous pass is insignificant because of relatively low temperature in this type of machining and due to high feeds used.

In hard machining, however, a unique combination of high temperatures at the tool–workpiece interface combined with low feeds is the case. This explains the so-called self-induced heat that reduces the cutting force and power in hard machining, constituting its essence. This became possible with the introduction of PCBN and ceramic tool materials that can withstand high temperatures. The formation of the white layer on the machined surface and high thermal residual stress constitute a price to pay for the discussed high temperatures.

The success of laser-assisted hard machining [32] provides an additional confirmation of the importance of the heat-affected zone. In operation, a laser beam is projected onto the part through fiber optics or some other optical-beam delivery unit, just ahead of the tool. The laser-induced heat softens the workpiece and makes it easier to cut. Because the laser beam is tightly focused, heating is localized around the actual cut. Heat is carried off in the chips and there are no changes in the physical properties of the material cut due to heat. Conventional CBN or ceramic cutting tools are used and have much longer life because the material cuts so easily.

1.3.2 Great Axial (Thrust) Components of the Cutting Force

Although Nakayama *et al.* [5] pointed out that the so-called springback of the work materials plays a significant role in hard machining, it was not noticed by subsequent researchers. Therefore a need is felt to clarify and quantify the issue.

When material is cut, the cutting tool deforms material first elastically and then plastically to separate the stock to be removed from the rest of the workpiece. Once the cutting edge passes a certain area of the surface being machined, the metal will spring back as the cutting load is released. In order to understand springback, it is necessary to consider the stress–strain diagram of the work material.

Scaled elongation–stress diagrams for annealed and quenched AISI 52100 steel are shown in Figure 1.7. First, consider the annealed diagram. When a force is applied, the work material deforms first elastically up to point A on the diagram. This point represents the so-called elastic limit. Within this limit, the work material is subjected to only elastic deformation so if the applied stress is released, the materials regains its initial size (point 0). The distance 0–B on the elongation axis represents the maximum elastic deformation. If the applied stress exceeds the elastic limit (the so-called yield strength, which is $\sigma_{y0.2} = 370\,\text{MPa}$ for annealed steel 52100), the material exhibits a combination of the elastic and plastic defor-

Figure 1.7 Elongation–stress diagrams for annealed and quenched AISI steel 52100

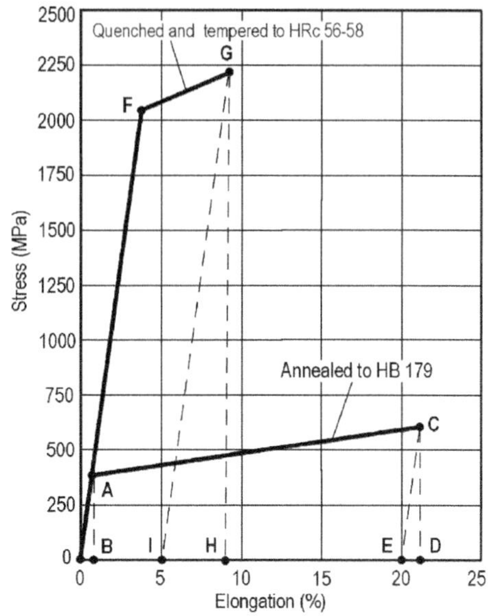

mations. The applied stress can grow further up to point C (corresponding to the so-called ultimate strength of the work material; for the considered case, $\sigma_{UTS} = 590$ MPa) on the diagram where fracture occurs.

The elongation corresponding to point C is known as the total elongation at fracture. In Figure 1.7, it is represented by distance 0–D on the elongation axis. After fracture, however, the applied stress is released and the permanent elongation found in the work material (represented by distance 0–E in Figure 1.7) which is known as the elongation δ. For the considered case, $\delta = 20\,\%$ is less than the total elongation at fracture by the elastic portion represented by distance E–D in the diagram. As such, the location of point E is readily found by drawing a line from point C parallel to line 0–A. As such, distance E–D is known as elastic recovery in materials testing or springback in materials processing. This springback causes the rubbing of the work material over the tool's flank face even if the flank wear land is not yet formed. As a wear land forms due to this rubbing, the tool–workpiece interface stresses increases significantly, causing flank wear.

Quenching does not change the modulus of elasticity E of the material. Thus, the elastic part of the elongation stress diagram follows the same line from point 0 to point F, which corresponds to the yield strength at the quenched state $\sigma_{y0.2} = 2030$ MPa. If stress is increase further to point G, which corresponds to the ultimate strength $\sigma_{UTS} = 2240$ MPa, the fracture occurs. In the diagram, distance 0–H represents the total elongation at fracture, 0–I represents the elongation that in the considered case is $\delta = 8\,\%$, and I–H is the elastic strain or springback.

Comparing springbacks of the work materials in the annealed and quenched conditions (distances E–D and I–H, respectively), one finds that the springback of the work material in the quenched stage is much greater. To the first approximation, the said springback can be determined as the ratio of the ultimate strength of the work material, σ_{UTS} and its elasticity modulus, E, *i.e.*, springback = σ_{UTS}/E. As the modulus of elasticity is almost the same for wide group of steels ($E = 200$ GPa), the springback is determined by the strength of the steel. For annealed AISI steel 52100 having $\sigma_{UTS} = 590$ MPa, springback = 0.00295, while for quenched having $\sigma_{UTS} = 2240$ MPa springback = 0.0112, *i.e.*, 3.8 times greater. This explains great radial (thrust) components of the cutting force and severe flank wear commonly found in hard machining. Residual heating, however, reduces this springback, while causing the formation of the white layer.

1.3.3 Great Power Spent in the Formation of New Surfaces

Cutting is different from other deformation problems in elasticity and plasticity, since after cutting, a single starting body has been separated into a number of entirely separate bodies that are no longer "attached" to the parent body. The work of separation is absent in traditional models of metal cutting because it was believed that it is insignificant. That view has been challenged by Atkins [12, 13, 15] who considers metal cutting as the branch of elastoplastic fracture mechanics that shows significance of the work for the formation of new surface in the energy balance of metal cutting. As this work increases with the strength of the work material, its relative impact in hard machining (Figure 1.6) becomes even more significant. Therefore, any model of chip formation in hard machining when it comes to the consideration of the force and/or energy balance must take into account the work needed for the formation of new surface. This is because the crack formation and propagation play an important role in hard machining.

1.3.4 Need for Rigid Machining Systems

As pointed out by Astakhov [26], the cyclic nature of chip formation in metal cutting is normally the major source of vibrations in metal cutting when other parameters of the cutting system are designed and made properly. When machining difficult-to-machine materials or even "normal" steels at high velocity, chip segmentation increases thus the variation of the cutting force within each cycle of chip formation increases proportionally. However, this variation of the cutting force normally does not present a problem in terms of machine dynamics unless severe seizure occurs on the tool–chip interface, which may cause some vibrations and even tool breakage.

In hard machining, no seizure at the tool–chip interface (between the chip and a PCBN tool, for example) has been reported so that conventional notions about "sticking–sliding" friction at the tool–chip interface is not applicable in this case to explain the highly dynamic nature of hard machining. Moreover, as pointed out by Nakayama *et al.* [5], the cutting force in hard machining is not high compared to the conventional machining when a sharp tool is used. Although it increases significantly with tool wear, low depths of cut and cutting feeds used in hard machining (relatively small volume to be removed) should not cause high cutting forces, thus should not present any dynamic problems. Therefore, another source of vibration in hard machining should be found and investigated in detail in order to improve the efficiency of this process.

As pointed out in the previous section, springback causing high contact stresses over the tool flank contact face is a feature of hard machining. In the author's opinion, the frictional vibration at the tool flank caused by these high stresses is the prime source of vibration. Frictional vibration is known as one kind of self-excited vibration which is defined as free vibration with negative damping. In self-excited vibration, the alternating force that sustains the motion is created or controlled by the motion itself; when the motion stops the alternating force disappears. Research on frictional behavior of materials is usually empirical in nature, since there is not yet a fundamental understanding of relevant frictional properties of materials. Part of the problem is that friction is not usually measured in a manner to determine potential vibration-induced mechanisms [33].

Published information on the frictional behavior of materials presumes the steady state and is not directly applicable to research on frictional vibrations, whereas the results of research on frictional behavior appear to show very different frictional properties which are not possible to verify by conventional friction tests. The major problem in the known approaches is that the driving force is assumed to be known or readily characterized. However, in most sliding systems, the driving force (variations in friction) is usually not well known, or must be derived from a simulative test. It is possible that frictional behavior of a material may change over a range of sliding speed and contact pressures to eliminate frictional vibrations, but this cannot be predicted from machine dynamics alone. At best then, frictional vibrations might be reduced to an acceptable amplitude by changes in system dynamics, or its frequency may be moved out of unacceptable ranges.

1.4 Basic Hard-machining Operations

Although hard machining can be used in practically any machining operation, hard-turning, hard-boring, hard-milling, and hard-gear-manufacturing operations have become most common. This section presents a short overview of these operations.

1.4.1 Hard Turning

The clear attraction to hard turning (Figure 1.8) is the possibility of eliminating grinding operations. However, for many shops, the process of repeatedly turning parts that are harder than 45 HRC to grinding-level accuracies is still unclear. Moreover, the economics of such a process is not well understood as efficiency of the process and cost per unit depend on many parameters, varying from one shop to another.

A properly "dialed-in" hard-turning process can deliver surface finish of R_a 0.4–0.8 μm, roundness of 2–5 μm, and diameter tolerance of ±3–7 μm. Such performance can be achieved on the same machine that "soft" turns the part prior to hardening, maximizing equipment utilization. However, some shops misstep by initially using the wrong (that is, less expensive) tool insert for the application. Others may not be sure if their machine possesses the rigidity to handle the highly dynamic thrust component of the cutting force that can be twice that of a typical turning operation.

Though a material of hardness 47 HRC is hard turning's starting point, hard turning is regularly performed on parts of hardness 60 HRC and even higher. Commonly hard-turned materials include tool, bearing, and case-hardened steels. Although Inconel, Hastelloy, Stellite, and other exotic materials are often considered as falling in the category of hard turning [34], it is not correct as their hardness is much less than 47 HRC and thus the mechanism of chip formation and process requirements including tool materials are considerably different.

From a metallurgical standpoint, materials with a small hardness deviation (less than two HRC points) throughout the cutting depth allow the best process predictability. Parts that are best suited for hard turning have a small L/D ratio and complicated profile. As mentioned above, in general, an L/D ratio for unsupported workpieces should be no more than 4:1. Despite tailstock support for long, thin parts, high cutting pressures would likely induce chatter.

The degree of machine rigidity dictates the degree of hard-turning accuracy. Most machines made in the last 10 years have sufficient rigidity to handle hard-turning applications. In many cases, a machine's overall condition is more of a factor than its age. Even an old, well-maintained manual lathe can be a candidate

(a) (b)

Figure 1.8 Hard turning: (a) high-speed, and (b) conventional

for hard turning. However, as required part tolerances get tighter and surface finishes get finer, machine rigidity becomes more of an issue. Special machines and turning centers made for hard machining are the best choice.

The key to success in hard turning, however, is the system rather than machine rigidity. Unfortunately, this simple rule is not well understood in industry. Maximizing system rigidity means minimizing all overhangs, tool extensions and part extensions, as well as eliminating shims and spacers. The goal is to keep everything as close to the turret or spindle head as possible.

As mentioned above, one of the prime challenges in the designing of a hard-turning process is whether or not to use the coolant. Cutting without coolant provides obvious cost benefits, thus most hard-turning operations are carried out dry. On the other hand, parts get hot, which makes process gauging and part handling difficult. Moreover, flying cherry-red chips may cause some additional problems. Therefore, if the use of coolant is needed, high-pressure through-the-tool coolant is the best choice to cool down the machined part and hot chips, keeping chip size small and shape easy to handle. As such, if a coolant is used, it must be water-based and of low concentration, for obvious reasons.

Because hard turning delivers the majority of cutting heat out in the chip, examining the chips during and after the cut will reveal whether or not the process is well-tuned. During a continuous cut, the chips should be blazing orange and flow off like a ribbon. If cooled chips essentially disintegrate when crunched by hand, then that demonstrates that the proper amount of heat is being produced in the chip.

The selection of proper tool material (PCBN, cermet, or ceramic) is vital for process efficiency and depends on the accuracy and surface finish required. The end users often are not aware that those listed are just generic types of tool materials. Within each type, hundreds of different grades are available from various tool-material, cutting-insert, and tool manufacturers. Therefore, the selection of proper tool-material grade is one of the most challenging tasks in hard turning in terms of obtaining efficient and stable machining process. The insert shape, tool holders and optimal machining regime just add more complications to this multi-variable optimization problem. Knowledge, understanding the essence of hard machining, and experience are prerequisites for success.

1.4.2 Hard Boring and Reaming

Boring, also called internal turning, and reaming are used to increase the inside diameter of an existing hole. The original hole is made with a drill, or it may be a cored hole in a casting. Boring and reaming achieve three prime objectives: (a) sizing – boring and reaming bring the hole to the proper diametric accuracy with a tight tolerance while achieving the required surface finish; (b) straightness – boring and reaming straighten the original drilled or cored hole; (c) concentricity – boring and reaming make the hole concentric with the outside diameter within the limits of the accuracy of the workholding fixture. This unique set of objectives is not

(a) (b)

Figure 1.9 (a) Common design of boring tool for hard machining, and (b) family of hole-finishing tools

normally achieved in grinding so it is logical to use hard boring and/or reaming as the finishing operation after the work material has been hardened.

If distortion and size variation as a result of heat treatment places unreasonable constraints on the "soft stage" machining, increasing its tooling cost, hard boring will provide a cost-effective, scrap-reducing alternative. The greater the distortion (due to part asymmetry, for example) and the length-to-diameter ratio, the greater advantages of hard boring.

Moreover, when properly used, hard boring and reaming is much more productive than internal grinding. According an example presented by a leading tool manufacturer, when internal grinding was replaced by hard boring in machining of hardened steel of hardness 63 HRC, the cycle time reduced from 26 min to 2 min 20 s. As such, the direct tooling cost and manufacturing costs related to the operation were reduced by 50 % and 100 %, respectively.

Typical boring and reaming tools for hard machining are shown in Figure 1.9. PSBN, cermet, or ceramic inserts are clamped into cartridges or directly to the tool body. To achieve high productivity, multi-insert tools are used. However, when high precision of the machined hole is required, single-blade tools with supporting pads are used. For bore sizes smaller than 12 mm in diameter, clamping of small inserts into small boring bars becomes progressively more difficult. This problem is solved by using brazed tools. The PCBN tip is brazed directly to a tungsten carbide shank, providing a rigid boring tool capable of producing good surface finishes and providing a highly productive alternative to internal grinding.

1.4.3 Hard Milling

Throughout the last few years, hard milling has captured the attention of manufacturers around the world. These manufacturers are typically focused on the mold and die industry where materials such as P20, H13, W5, S7, and others are commonly cut. Traditionally, core and cavities from these materials are manufactured in the hardened state using electrical-discharge machining. Through the years, new technologies have been developed where these materials can be, in most cases, machined directly into hardened material using new toolpath processing techniques to form hard milling. These materials can range from 45 HRC to as hard as 64 HRC. Advanced moldmakers have realized that adopting new technology can be one of their keys to survival against global competition.

According to Zurek [35], successful hard milling is the result of implementing a system including the machine, cutting tools and toolholders, and the computer-aided design/manufacturing system. The machine tool is the key component of the system. The machine must be designed for hard milling, along with having some of the same characteristics found in a high-speed machining center. The base construction and the individual components of the machine, such as the drive train, spindle, and CNC system, must be able to handle the demands of hard milling. A rigid base with good vibration damping characteristics is of prime importance. Polymer concrete bases are a good choice for high-speed and hard milling applications because they typically have vibration damping 6–10 times greater that of cast iron.

Digital drives that can handle fast acceleration/deceleration provide good contouring accuracy while helping to minimize cutting-tool wear. Spindles should provide flexibility, offering high torque at low speeds and high power over a large speed range.

Mold shops use three general types of hard milling tools: solid carbide endmills, indexable carbide inserts, and, most recently, ceramic indexable inserts. Each of these tools has its strengths and weaknesses depending upon the application. Solid carbide endmills are usually precision ground, coated, and quite expensive. The second type of hard milling tool is a cutter with indexable carbide inserts. In most cases the carbide grades and geometry of these inserts are not designed well for hard milling, and they do not offer optimal tool life or productivity in hardened materials. The third type is ceramic indexable inserts, more specifically, whisker-reinforced ceramic inserts. The benefits of using a system of cutters with indexable ceramic inserts include faster cycle times and a reduced number of operations per part. A full line of cutters for hard milling with whisker-reinforced ceramics enables a shop to rough out a part from a solid hardened block – including face milling, pocketing and profiling with indexable inserts – and finish it in one setup. Cutters engineered to mill with ceramics are capable of secure, high-speed milling from large face mills down to small diameter endmills – all using indexable ceramic inserts. It is important to use cutters designed for hard milling with ceramics at high velocity for secure insert clamping. Modern whisker-reinforced ceramics have a melting point of more than 2000 °C, which means that ceramic inserts can operate at speeds well beyond the point where carbide tools fail. In fact, whisker-reinforced ceramics work better above the melting temperature of carbide inserts. Coolant is not recommended for hard milling applications with ceramic inserts, but air blast is suggested especially when pocket milling to keep from recutting chips. Reduced coolant usage and disposal cost is an added benefit when using ceramic inserts for hard milling.

Toolholders play an important role in hard milling. Because hard milling requires a large range of rotational speeds (from low for a roughing application to high for finishing), the hollow-shank tooling interface between the toolholder and spindle interface should be used. It provides rigid and balanced tooling setup over the ISO taper interface. Although collet chucks and hydraulic expansion toolholders are an excellent choice for roughing and semi-finishing operations, for finishing hardened cavities and cores with a high degree of accuracy and quality, power and heat-shrink toolholders provide excellent characteristics. Today, all systems are commonly available from most tooling suppliers.

1.4.4 Hard Broaching

In the field of metal cutting for mass production of parts with complex profiles, the broaching operation is the most economical method if high production rates combined with great consistency of machined parts are required. The advantages of broaching are based on its technical principle which includes a multi-toothed tool with cutting edges one after the other and graduated in depth of chip thickness. The profile of a part can be broached in single stroke. Internal broaching is started from a pre-machined hole, while external broaching is to machine a surface profile. Broaching is possible in both directions horizontal and vertical. Cutting motion can be linear or helical.

Two different methods of hard broaching are feasible with this tool configuration:

- hard broaching without defined stock removal: the parts are finish-broached before hardening and hard broaching means only clearing the heat distortion;
- hard broaching with defined stock removal (0.1–0.2 mm diametrical), which requires a corresponding finish of the pre-broaching tool considering the expected heat distortion.

It is known that the cutting speed in common broaching operations is in the range of 0.5–6 m/min. This speed is 10–30 times less than that in turning and milling of the same work material. Among many reasons for these low speeds, the common tool material, high-speed steel, is the root cause. The advent of hard broaching gave rise to the use of more advanced tool materials such as carbides and PCBN. Since then, the concerns about productivity, quality of the machined surface, and the efficiency and reliability of hard broaching came to the attention of researchers and practitioners.

The following summarize the results obtained so far:

- Coated carbide broaches designed as a flank cutting tool show the best results in terms of quality and reliability of broaching operations. Profile quality IT 7–8 can be achieved. Surface finish is normally in the range of $R_a = 0.3$–$0.5\,\mu$m and is dependent on the combination of the carbide grade and coating as well as the adhesion properties of the work material.

- An effective cutting speed of 60–63 m/min is determined as an important hard-broaching process parameter when carbide broaches are used. By this speed, the broaching stroke lasts 1–2 s, implicating shorter cycle times in comparison to green (soft) broaching. In some operations, as for example, internal gear manufacturing or spline broaching of sun gears for automotive transmissions, hard broaching is the only feasible and economical way to improve quality of ring and sun gears after their heat treatment.
- Smaller uncut chip thickness t_1 (the rise per tooth) compared to conventional broaching is the case. As such, the tool life characterized by the intensity of tool wear (the linear wear rate v_L) [36] is proportional to t_1, as reported by Makarov *et al.* [37] in machining of high-nickel (35%) alloy, when $t_1 = 0.02$ mm, $v_L = 2$ μm/m, $t_1 = 0.06$ mm, $v_L = 3$ μm/m, $t_1 = 0.10$ mm, $v_L = 3.5$ μm/m.
- PCBN can also be used as the tool material in hard broaching, allowing higher cutting speeds (up to 100 m/min) compared to carbide broaches [38]. However, the process is unstable in terms of cutting-edge chipping after only few meters of machining, which was attributed to the quality of PCBN with no reason given.
- Hard broaching is only possible on specially designed machines as ordinary broaching machine cannot deliver the high cutting speed and cannot withstand excessive broaching axial force.

1.4.5 Hard-gear-manufacturing Operations

The function of gears remains an important element in modern machinery. Gear manufacturers face numerous challenges to improve their production capabilities in today's competitive markets. They must improve efficiency to meet customer demands for gear sets with lower vibration, lower noise in wide frequency range, and increased load ability. Customers also require increased gear hardness and improved tooth surface quality and accuracy. While addressing these demands, gear manufacturers also remain concerned with the environmental impact of using oil products or other pollutants. Reducing the use of cutting fluids, or completely removing the coolant by dry cutting, must be considered in the production process. To meet all these challenges, gear manufacturers must increase their production capacity and reduce costs through the use of precision cutting techniques.

The mentioned requirements for gears result in the implementation of special alloys for gears and their heat treatment to achieve the required strength and hardness which are requirements for gear accuracy, low noise, and durability. To achieve accuracy requirements in machining of hard gears, additional operations for gear finishing are required. Normally, it is necessary to have/purchase expensive specialized equipment for gear-finishing operations such as grinding and honing, applied after gear heat treatment to achieve the required quality. These operations increase the cost of gears significantly. As an alternative, hard-gear-cutting operations such as hard hobbing, shaving, and broaching can be used.

Hard hobbing is able to do both rough machining before heat treatment and finishing work after heat treatment. It reduces the cost of gears, increases the produc-

(a) (b)

Figure 1.10 (a) Solid carbide hob, and (b) double-start indexable carbide insert gear hob

tion rate, and allows centralization of equipment. Hard hobbing, however, sets some special requirements on the tool, the machine, and the operation as a whole. Understanding these requirements results in successful implementation of this process.

The most common hob used today in industry is made of high-speed steel. Because this tool material does not possess sufficient wear resistance and red hardness,[1] its use has been limited to the machining of relatively soft work materials having Brinell hardness of no more than 250. Further, a hob made of high-speed steel cannot finish up the surface of even such relatively low hardness materials but can machine it only to an extent as to require further finishing work by other gear manufacturing tools, for example, shavers. The major reason is an insufficient flank angle along the teeth profile.

Generally in manufacturing, it is customary to form the cutting wedges of form-cutting tools with a flank angle of 3–4° to improve the properties of a finished surface [19]. In hobbing tools, the flank angles used are much smaller because it is necessary to keep the tooth profile within a narrow tolerance on hob multiple resharpenings. In machining of hard materials, the springback of the work materials is great due to a high yield strength, so excessive rubbing on the flank surface is the case. High temperatures occurring due to this rubbing cause adhesion between the flange and the work material that ruins the surface finish and reduces tool life. Therefore, tool materials that can withstand much higher contact temperatures and prevent high-temperature adhesion with the work materials are used for hard hobbing.

Sintered carbides are common tool materials used in hard hobbing. Figure 1.10 shows a solid carbide hob used for small modules and an indexable carbide insert gear hob. The latter shows the maximum advantage in hard hobbing although it is an expensive tool that requires a long pre-setting time and proper handling. Although the advantages and potentially efficiency gains in using of cermet and PCBN hobs have been revealed, the practical use of these tool materials in hard hobbing requires additional studies, and new machines and tooling designs.

[1] The property of being hard enough to cut metals even when heated to a dull-red color.

1.4.6 Skiving

1.4.6.1 Hard "Skiving" Turning

Traditionally when cutting tools with polycrystalline superhard materials (PSHM) are used, the cutting feed does not exceed 0.2 mm/rev due to a common requirement of obtaining surfaces with roughness of R_a and R_z below 1.25 and 5 μm, respectively (Figure 1.11) [39]. Increased turning efficiency is possible by using cutting tools with wiper geometry [40], tools with non-planar (sculptural) rake faces [41], and "skiving" cutting [39].

The working principles and advantages of cutting tools with wiper geometry and with sculptural rake faces in terms of improving allowable feed are well known [42] while there is very little information about "skiving" machining with tools equipped with CBN-based PSHM. The results of a study on hard turning of high-chromium–nickel alloy (hardness of 60–62 HRC) were discussed by Klimenko and Manokhin [39]. As pointed out, surface roughness of R_a = 0.07–0.14 μm was achived at feed of 0.05 mm/rev. In machining the SCM415 steel with a hardness of 61 HRC using Sumitomo PCBN-based tool material, the surface roughness R_z were 1.25 and 1.75 μm at feeds 0.1–0.3 mm/rev, respectively [13].

As discussed by Astakhov [19], the proper determination of the uncut (undeformed) chip thickness is of prime importance in the optimization of any machining operation and in the determining the energy balance. Due to "skiving" turning special features, the determination of the cross-section parameters of the uncut chip thickness in "skiving" turning is of particular interest as it is the prime parameter to calculate the chip compression ratio and thus the energy spent in plastic deformation of the work material as discussed above.

The model of "skiving" turning proposed by Klimenko and Manokhin [39] is shown in Figure 1.13. In this model, region 1–2 of the chip is the line of intersection of the plane, which coincides with the tool rake face having a zero rake angle with the surface (machined at the previous turn) that represents a hyper-

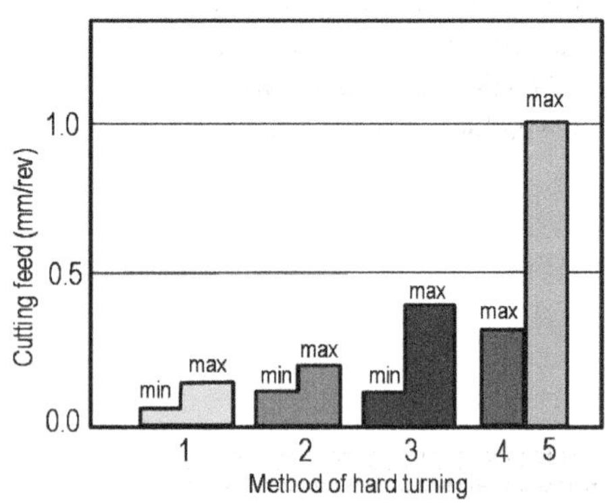

Figure 1.11 Minimum and maximum feeds used in various hard-turning methods: standard tools (f = 0.05–0.12 mm/rev) (1), turning with tools having a cylindrical rake face (f = 0.1–0.2 mm/rev) (2), turning with tools having a wiper geometry (f = 0.1–0.4 mm/rev) (3), traditional "skiving" turning (f = 0.3 mm/rev) (4), and advanced "skiving" turning (f up to 1.0 mm/rev) (5)

Figure 1.12 Hard "skiving" turning

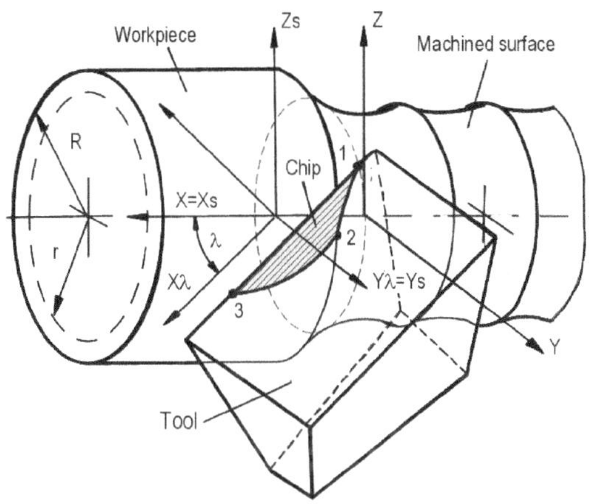

Figure 1.13 Model for the determination of the uncut chip thickness in "skiving" turning

boloid of rotation. Region 2–3 is the line of intersection of surface to be machined on the workpiece (a cylinder) with the rake face. The shape of the chip cross-section is bounded by the tool cutting edge and curve 1–2–3.

The uncut chip thickness t_1 in each point of the cutting edge is understood as a distance between the cutting edge on the transient surface and line 1–2–3 measured along the normal to the tool cutting edge.

$$t_1(X_\lambda) = \begin{cases} \sqrt{r^2 - X_\lambda^2 \sin^2\lambda + (f + X_\lambda \cos\lambda)^2 \tan\lambda} \text{ for } X_1 \le X_\lambda < X_2 \\ \sqrt{R^2 - X_\lambda^2 \sin^2\lambda} \text{ for } X_2 \le X_\lambda \le X_3 \end{cases} \quad (1.4)$$

where

$$X_1 = \frac{-f}{2\cos\lambda},\ X_2 = \frac{\sqrt{R^2 - r^2}}{\sin\lambda} - \frac{f}{\cos\lambda},\ X_3 = \sqrt{\frac{R^2 - r^2}{\sin^2\lambda}} \quad (1.5)$$

In these equations, r is the radius of the machined surface, R is the radius of the workpiece, f is the cutting feed, and λ is the cutting-edge inclinational angle [19, 43].

Figure 1.14 shows dependences of the average uncut chip thickness on various machining parameters. As seen, the average chip thickness depends on the diameter of the workpiece, cutting-edge inclination angle, λ, depth of cut, d_w, and cutting feed, f. In the range of machining parameters $f = 0.5$–2.0 mm/rev, $d_w = 0.05$–0.15 mm, and $\lambda = 15$–$60°$, the average chip thickness is in the range of 10–60 μm. The results obtained show that feed and cutting-edge inclination have the strongest effect on the average uncut chip thickness (Figure 1.14 (a)) and with an increase of one of these characteristics the degree of the influence of the other increases. In comparison with the above characteristics, an increase in the cutting depth increases the average uncut chip thickness to a lesser degree (Figure 1.14 (b)) and as the diameter of the part machined increases, the chip thickness decreases.

Summarizing the experimental results [39], the following important features of "skiving" hard turning (compared to usual hard turning with PCBN tools) can be represented as:

- High productivity and efficiency: the cutting speed in "skiving" turning is the same while the cutting feed is up to five times greater.
  - Long tool life: the average uncut chip thickness is relatively small even when the cutting feed (known in industry as the chip load) reaches 1 mm/rev (Figure 1.14). This assures lower machining temperatures and uniform wear of the rake and flank tool contact areas as shown in Figure 1.15.

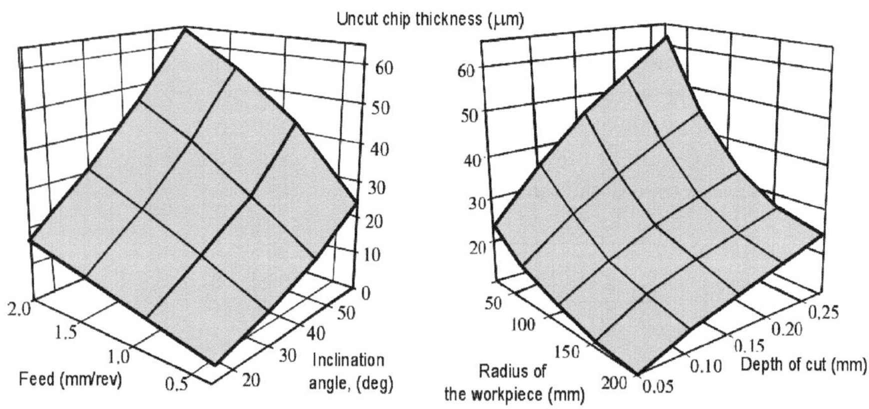

Figure 1.14 Effect of the machining parameters on the average uncut chip thickness: (a) $\boldsymbol{R} = 50$ mm, $\boldsymbol{t} = 0.1$ mm, and (b) $\boldsymbol{f} = 1$ mm/rev, $\lambda = 45°$

- The most distinguishing feature of "skiving" turning is that the roughness of the machined surface does not deteriorate proportionally to the tool wear as in the conventional hard machining (Figure 1.2). Figure 1.16 shows an example. As seen, if the tool life criteria is accepted to be $VB_B = 0.25$ mm then practically no surface finish deterioration occurs over the entire period of tool life.
- High quality of the machined surface: the depth of the "white" layer is in the range from 5 to 16 μm. No solid phase transformation due to high machining temperature was found.
- Surface roughness is much smaller compared to usual hard turning, while the feed is up to five-fold higher. Figure 1.17 shows an example of the experimental data [39]. At cutting speeds in the range of 0.9–1.2 m/s, the roughness of the machined surface depends on the cutting feed as: at $f = 0.09$–0.67 mm/rev, $R_a = 0.3$–0.6 μm, $R_z = 1.5$–6 μm; at $f = 0.67$–1.30 mm/rev, $R_a = 0.6$–0.8 μm, $R_z = 4$–9 μm; at $f = 1.30$–2.60 mm/rev, $R_a = 0.80$–1.25 μm, $R_z = 6$–12 μm.

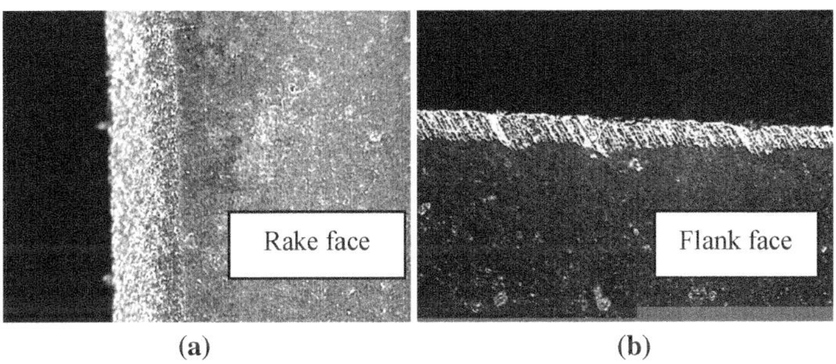

Figure 1.15 (a) Rake and (b) flank faces having uniform wear pattern

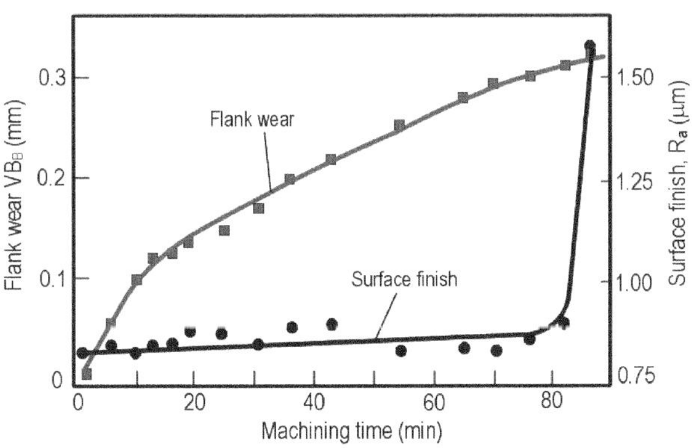

Figure 1.16 Roughness of the machined surface variation with the tool flank wear in machining AISI E52100 (0.98–1.1 % C, 1.45 % Cr, 0.35 % Mn) of hardness 60–62 HRC with $f = 0.67$ mm/rev, $v = 1.32$ m/s, $\lambda = 50°$

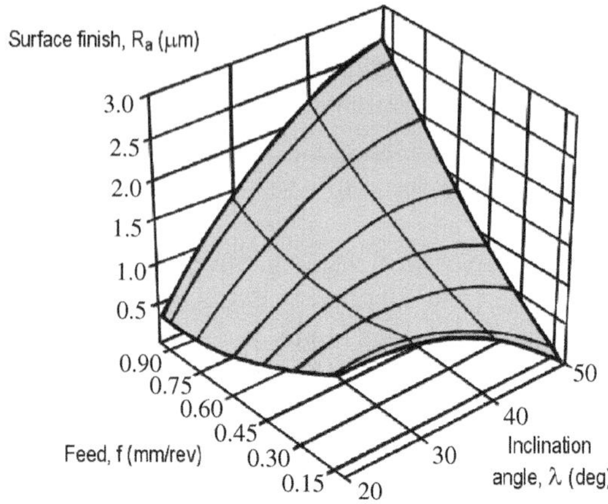

Figure 1.17 Peak-to-valley height R_a *vs.* feed and cutting-edge inclination angle. Workpiece material AISI 4340, 48–50 HRC, $d_w = 0.15$ mm, $v = 2.5$ m/s

1.4.6.2 Gear Skiving

Skive hobbing is finish hobbing of hardened gears. This process removes the inaccuracies in the gear flanks and profile which occur during heat treatment. The result is improved gear accuracy. It is possible to hob hardened gears in the range up to 63 HRC using tungsten carbide hobs. This gear-finishing process is capable of achieving gear quality levels of up to DIN class 6 or AGMA class 11 in standard module ranges of 1 to 6 mm. Gear skiving has gained in recent years a considerable reputation among gear producers and is nowadays a powerful alternative to traditional grinding and hard hobbing [44]. This became possible with the development of highly evolved and automated machine tools, which make the method practical and economical [45]. Moreover, the introduction of well-designed cemented carbide tools and the implementation of stock-dividing systems in gear-hobbing machine tools make the skiving process attractive and efficient.

Figure 1.18 shows Gleason's skive hobbing of a module 16 mm hardened spur gear using a Gleason P 600 hobbing machine. The hob tool consists of carbide cutting inserts brazed on a high-speed steel body. The skiving process removes in one cut 0.5 mm stock per gear flank. Besides significant reduction of tooling cost and cycle time, advantages of this process include capabilities such as gear flank modifications like taper and crown in the range of 100 μm and more in only a single cut. DIN 8 quality was achieved on the profile, which is the most sensitive quality characteristic in this process.

The special geometry of the skiving hob teeth is shown in Figure 1.19. As it appears for many researchers, the main differences between skiving and gear hobbing cutting teeth are the negative tool-in-hand rake angle γ_k formed due to the tooth rake offset δ_k [44]. It appears that the negative rake angle protects the carbide cutter from shocks and instantaneous overloading.

In reality, however, it is not quite the case as the tool-in-machine tooth geometry [19] is not considered, so the process modeling is normally overcomplicated and essential variables are left out of consideration. If the tool-in-machine tooth geometry is considered, the role of the cutting-edge inclination angle is revealed and the results discussed in the previous section on "skiving" turning are fully applicable in skiving hobbing.

Figure 1.18 Skiving hobbing

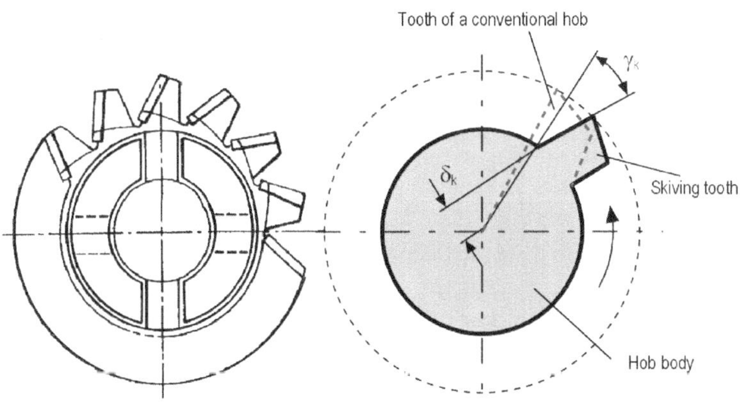

Figure 1.19 Comparison of the tooth geometry used in conventional and skiving hobs

1.4.7 Hard Machining with a Rotating Cutting Edge

There are two rationales behind this type of hard machining:

1. As is well known [26, 29, 46], the so-called kinematical surface roughness (its parameter R_z in particular) is much smaller when a round insert or an insert with the nose radius larger than the depth of cut is used. As such, $R_z \approx f^2/(8r_n)$, where f is the cutting feed, and r_n is the insert or tool nose radius. This fact is used in hard machining (turning, drilling, and milling) to achieve fine surface roughness required in finishing operations.
2. Hard machining has become an option with the appearance of improved tool materials such as CBN, PCBN, and ceramics. The CBN or PCBN tool materials are very expensive, while the latter, hard-cutting ceramics, have a much shorter tool life, but a much lower cost. The major problem causing lower tool life of ceramics and carbide tool materials is highly concentrated heat generation in hard machining that causes great temperature gradients and concentrated tool wear. If the cutting tool is designed so that a fresh cutting-edge portion is moved into position to replace the portion of the cutting edge being used the it would be possible to use the less expensive ceramic tool materials, superior grades of carbide, or other low-cost tool materials capable of hard machining, provided the non-productive tool changing time could be reduced and the tool material could be used more efficiently.

The simplest realization of the concept is proposed by Shaw (US Patent No. 6,733,365, 2004). According to this patent, the same tool can be used for combining a comparatively coarse, roughing cut at high removal rate and a finishing cut at a low removal rate commonly used in hard-machining operations. It is accomplished by rotating the cutting inserts between roughing and finishing so that a fresh portion of the cutting edge is used for the finishing pass. Figure 1.20 shows a detail of the proposed tool. A means for rapid and precise indexing and prevention of rotation of the insert during cutting is provided by a gear attached to the bottom portion of the axis on which the insert is located. An adjustable step motor actuates the worm gear set.

Another realization of the discussed concept is conventional rotary tools. The continuous spinning of the cutting insert about its axis in addition to the main cutting and feed motions is the major difference between rotary cutting and conventional cutting [47–49]. Figure 1.21 shows the principle of such a tool. As seen, in addition to the prime motion with the rotation speed n_w and feed motion, f, the cutting insert is provided with rotation n_{in} about its axis. As a result, the cutting edge moves continuously so that fresh portions of this edge are entering into the machining zone.

Initially, rotary tools were developed so that the rotation of the cutting insert was provided by the forces acting in cutting. Such tools became known as self-propelled rotary tools. The major advantage of these tools is their simple design

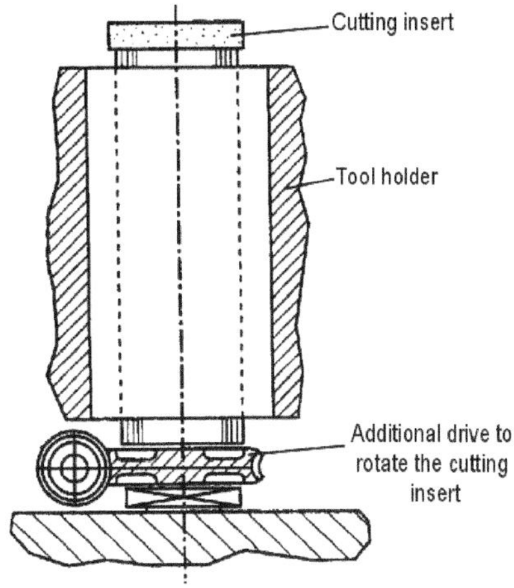

Figure 1.20 Cutting tool according to US Patent No. 6,733,365, 2004

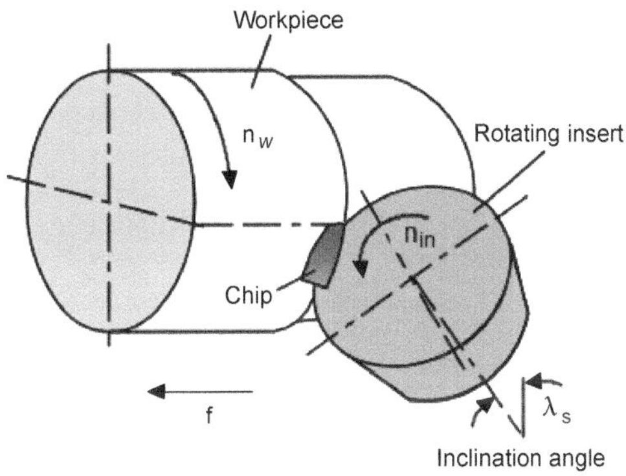

Figure 1.21 Kinematics of rotary turning

and versatility. The major disadvantage is instability, cutting speed, and feed-dependent rotation of the cutting insert due to many variables influencing the forces involved. The optimal cutting geometry is achieved by adjusting the proper direction of the forces to assure the rotation of the cutting insert. Although for many years research efforts were undertaken in many countries, this tool concept has never found its way into practical production besides a few isolated cases as the balance between the conditions of reliable cutting-insert rotation and the tool geometry (to gain advantages of such a design) exists only in a rather narrow range of process parameters so that it cannot be assured in many practical applications. Examples of self-propelled single-point and milling tools are shown in Figure 1.22.

To overcome the listed problems in self-propelled rotary tools, spinning tools were developed where the insert is rotated by an independent external source, *e.g.*, an electric motor. In such tools, the rotation of the cutting insert is stable and it does not depend on the machining regime, properties of work and tool

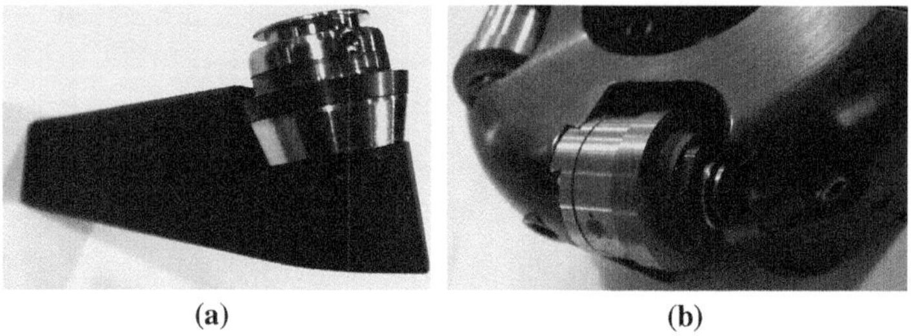

Figure 1.22 (a) Self-propelled single-point tool, and (b) milling tool

Figure 1.23 Spinning tool jointly developed by Mori Seiki and Kennametal

materials, and other variables of a particular machining operation. Moreover, with the development of the spinning-tool design, the tool geometry can be set to achieve the optimal tool performance. Figure 1.23 shows the spinning tool jointly developed by Mori Seiki and Kennametal companies.

In rotary tools, the rotation of the cutting insert allows continuous changing of the cutting edge involved in cutting so that each portion of circumference of the insert is engaged in cutting for very short time period. It allows increase of the material removal rate restricted by the high cutting temperature in conventional turning with single-point tools. As a result, the productivity of machining and tool life are increased. This was the rationale behind the design of any rotary tool. What was noticed, however, is the great differences in tool geometry between self-propelled and spinning rotary tools. The major difference is in the inclination angle (Figure 1.22), which normally reaches 30–40°. Thus the full advantage of the above-discussed "skiving" turning is realized, i.e., a preferable state of stress in the deformation zone is formed as the tool rake angle becomes close to the effective rake angle [19]. As such, the chip deformation is small due to the small work of plastic deformation of the work material.

References

[1] Richt C (2009) Hard turn toward efficiency. Gear Solutions (4): 22–30.
[2] Chou YK, Evans CJ (1999) White layers and thermal modeling of hard turned surfaces. International Journal of Machine Tools and Manufacture 39(12): 1863–1881.
[3] Hashimoto F, Guo YB, Warren AW (2006) Surface integrity difference between hard turned and ground surfaces and its impact on fatigue life. CIRP Annals – Manufacturing Technology 55(1): 81–84.
[4] Bosheh SS, Mativenga PT (2006) White layer formation in hard turning of H13 tool steel at high cutting speeds using CBN. International Journal of Machine Tools and Manufacture 46(2): 225–233.
[5] Nakayama K, Arai M, Kanda T (1988) Machining characteristic of hard materials. CIRP Annals – Manufacturing Technology 37: 89–92.
[6] Shaw MC, Vyas A (1993) Chip Formation in the Machining of Hardened Steel. CIRP Annals – Manufacturing Technology 42(1): 29–33.
[7] Tönshoff HK, Wobker HG, Brandt D (1995) Tribological aspects of hard turning with ceramic tools. Journal of Society of Tribologists and Lubrification Engineers 51(2): 163–168.
[8] Tönshoff HK, Arendt C, Ben Amor R (2000) Cutting of hardened steel. CIRP Annals – Manufacturing Technology 49(2): 547–566.
[9] Davies MA, Burns TJ, Evans CJ (1997) On the dynamics of chip formation in machining hard metals. CIRP Annals – Manufacturing Technology 46(1): 25–30.
[10] Elbestawi MA, Srivastawa SAK, El-Wardany TI, (1996) Model for chip formation during machining of hardened steel. CIRP Annals – Manufacturing Technology 45(1): 71–76.
[11] König W, Klinger M, Link R (1990) Machining hard materials with geometrically defined cutting edges. CIRP Annals – Manufacturing Technology 39(1): 61–64.
[12] Atkins AG (2006) Toughness and oblique metalcutting. ASME Journal of Manufacturing Science and Engineering 128: 775–786.
[13] Atkins T (2009) The Science and Engineering of Cutting, Amsterdam: Butterworth-Heinemann.
[14] Atkins AG, Mai YW (1985) Elastic and Plastic Fracture: Metals, Polymers. Ceramics, Composites, Biological Materials, New York: John Wiley & Sons.
[15] Atkins AG (2003) Modelling metal cutting using modern ductile fracture mechanics: qualitative explanations for some longstanding problems. International Journal of Mechanical Science 45: 373–396.
[16] Huang Y, Liang SY (2003) Cutting forces modeling considering the effect of tool thermal property – application to CBN hard turning. International Journal of Machine Tools and Manufacture 43(3): 307–315.
[17] Lalwani DI, Mehta NK, Jain PK (2008) Experimental investigations of cutting parameters influence on cutting forces and surface roughness in finish hard turning of MDN250 steel. Journal of Materials Processing Technology 206(1–3): 167–179.
[18] Remadna M, Rigal JF (2006) Evolution during time of tool wear and cutting forces in the case of hard turning with CBN inserts. Journal of Materials Processing Technology 178(1–3): 67–75.
[19] Astakhov VP (2010) Geometry of Single-Point Turning Tools and Drills. Fundamentals and Practical Applications, London: Springer.
[20] Astakhov VP, Xiao X (2008) A methodology for practical cutting force evaluation based on the energy spent in the cutting system. Machining Science and Technology 12: 325–347.
[21] Astakhov VP, Outeiro JC (2008) Metal cutting mechanics, finite element modeling. In: Davim PJ (ed) Machining fundamentals and recent advances. Springer, London, pp 1–27.
[22] Incopera FP, De Witt DP, (2001) Fundamentals of Heat and Mass Transfer. 5th ed., New York: John Wiley and Sons.
[23] Manca O, Morrone D, Nardini S (1999) Thermal analysis of solids at high peclet numbers subjected to moving heat sources. ASME Journal of Heat Transfer 121–1: 182–186.

[24] Muzychka YS, Yovanovich MM (2001) Thermal resistance models for non-circular moving heat sources on a half space. Journal of Heat Transfer 123(4): 624–632.
[25] Ostafiev VA, Noshchenko AN (1985) Numerical analysis of three-dimensional heat exchange in oblique cutting. CIRP Annals-Manufacturing Technology 34(1): 137–140.
[26] Astakhov VP (1998/1999) Metal Cutting Mechanics, Boca Raton, USA: CRC Press.
[27] Astakhov VP, Shvets S (2004) The assessment of plastic deformation in metal cutting. Journal of Materials Processing Technology 146: 193–202.
[28] Astakhov VP, Shvets SV (2001) A novel approach to operating force evaluation in high strain rate metal-deforming technological processes. Journal of Materials Processing Technology 117(1–2): 226–237.
[29] Astakhov VP (2006) Tribology of Metal Cutting. Triblogy and Interface Engineering Series, No. 52, ed. B.J. Briscoe, London: Elsevier.
[30] Poulachon G, Moisan A, Jawahir IS (2001) On modelling the influence of thermo-mechanical behavior in chip formation during hard turning of 100Cr6 bearing steel. CIRP Annals – Manufacturing Technology 50(1): 31–36.
[31] Silin SS (1979) Similarity Methods in Metal Cutting (in Russian). Mashinostroenie, Moscow.
[32] Dumitrescu PK, Koshy P, Stenekes J, Elbestawi MA (2006) High-power diode laser assisted hard turning of AISI D2 tool steel. International Journal of Machine Tools and Manufacture 46(15): 2009–2016.
[33] Ludema KC (1996) Friction, Wear, and Lubrication – a Textbook in Tribology, Boca Raton (Florida): CRC Press.
[34] Grzesik G (2008) Machining of hard materials, in Machining: Fundamentals and Recent Advances, P. Davim, Editor, Springer: London. p. 97–126.
[35] Zurek G (2004) The secrets to hard milling success. Moldmaking Technology (4): 14–18.
[36] Astakhov VP (2004) The assessment of cutting tool wear. International Journal of Machine Tools and Manufacture 44: 637–647.
[37] Makarov VF, Tokarev DI, Tyktamishev VR (2008) High speed broaching of hard machining materials. International Journal of Material Forming 1(1): 547–550.
[38] Klocke, F., Brinksmeier, E. and Weinert, K. (2005) Capability profile of hard cutting and grinding processes. CIRP Annals-Manufacturing Technology 54(2): 552–580.
[39] Klimenko SA, Manokhin AS (2009) Hard "skiving" turning. Journal of Superhard Materials 31(1): 42–54.
[40] Grzesik W, Rech J, Wanata T (2007) Surface finish on hardened bearing steel parts produced by superhard and abrasive tools. Int. J. Machine Tools & Manufacture 47: 255–262.
[41] Klimenko SA (1998) Special features of machining protective coatings. Journal of Superhard Materials 20(3): 440–445.
[42] Astakhov VP (2010) Geometry of Single-Point Turning Tools and Drills. Fundamentals and Practical Applications, London: Springer.
[43] Astakhov VP, Davim PJ (2008) Tools (geometry and material) and tool wear. In: Davim PJ (ed) Machining: fundamentals and recent advances. Springer, London, pp 29–58.
[44] Antoniadis A, Vidakis N, Bilalis N (2004) A simulation model of gear skiving. Journal of Materials Processing Technology 146: 213–220.
[45] Gimpert D (1991) Fine pitch gear hobbing advances. SME Technical Paper MS91-246.
[46] Shaw MC (2004) Metal Cutting Principles. 2nd Edition, Oxford: Oxford University Press.
[47] Dessoly V, Melkote SN, Lescalier C (2004) Modeling and verification of cutting tool temperatures in rotary tool turning of hardened steel. International Journal of Machine Tools and Manufacture 44(14): 1463–1470.
[48] Diniz AE, Ferreira JR, Filho FT (2003) Influence of refrigeration/lubrication condition on SAE 52100 hardened steel turning at several cutting speeds. International Journal of Machine Tools and Manufacture 43(3): 317–326.
[49] Kishawy HA, Wilcox J (2003) Tool wear and chip formation during hard turning with self-propelled rotary tools. International Journal of Machine Tools and Manufacture 43(3): 433–439.

Chapter 2
Advanced Cutting Tools

L.N. López de Lacalle, A. Lamikiz, J. Fernández de Larrinoa and I. Azkona

In this chapter the basic design principles and the current state-of-the-art for cutting tools specially designed to be applied on difficult-to-cut materials are described. One by one, the main aspects involved in tool design and construction will be explained in depth over the following sections, completing a general view of the tool world, to provide easy comprehension of the whole book. Materials for the substrates, coatings, and geometry are explained, with special attention to recent developments. A section is devoted to new machining techniques such as high-feed and plunge milling, turn milling and trochoidal milling.

2.1 Materials for Cutting-tool Manufacture

Cutting tools must simultaneously withstand big mechanical loads and high temperatures. Temperature in the chip/tool interface reaches more than 700 °C in some

L.N. López de Lacalle
Department of Mechanical Engineering, University of the Basque Country,
Faculty of Engineering of Bilbao, c/Alameda de Urquijo s/n, E-48013 Bilbao, Spain
e-mail: norberto.lzlacalle@ehu.es

A. Lamikiz
Department of Mechanical Engineering, University of the Basque Country,
Faculty of Engineering of Bilbao, c/Alameda de Urquijo s/n, E-48013 Bilbao, Spain
e-mail: aitzol.lamikiz@ehu.es

J. Fernández de Larrinoa
Metal Estalki, Derio, Bizkaia, Spain
e-mail: jflarrinoa@metalestalki.com

I. Azkona
Metal Estalki, Derio, Bizkaia, Spain
e-mail: iazkona@metalestalki.com

cases. Additionally, the friction between tool and removed chip, on one hand, and tool against the new machined surface, on the other, is very severe. Bearing this in mind, the main factors for a good tool design and post-manufacturing are:

- Cutting-tool substrate material must be very stable chemically and physically at high temperatures.
- Material hardness must be kept to the high temperatures suffered at the chip/tool interface.
- Tool material has to present a low wear ratio, both for the abrasion and adhesion mechanisms.
- Tool material must present enough toughness to avoid fracture, especially when operation to perform implies interrupted or intermittent cutting.

In the following sections each of the main tool materials are going to be described, starting from the lowest hardness to the highest. These groups are:

- *High-speed steels* (HSS), including the new powder-sintered grades. However, this material family has not enough hardness for hard machining.
- Sintered carbides, usually known as *hardmetal.* They are a compound of submicron tungsten carbide grains with a binder (usually cobalt, 6–12 %) This kind of material in the straight grade or in the coated grades (see an example in Figure 2.1) is the most used today for hard machining and high-speed machining.
- Ceramics based on alumina (Al_2O_3) or silicon nitride (Si_3N_4).
- Extra-hard materials, *i.e.,* polycrystalline diamond (PCD) and polycrystalline cubic boron nitride (PCBN), in different grades.

Before explaining the main aspects of each material, a mention of the company type involved in tool fabrication is an interesting point. Thus, in the current tool market two types of company are possible: firstly, the producers of basic tool materials, usually big international companies such as CeraTizit, Krupp, Sumitomo, General Electric, De Beers, Sandvik, Kennametal, Iscar and others, which also manufacture the complete cutting-tool systems including toolholders, inserts or integral cutting tools. Currently these companies represent the 80 % of the total world market.

Figure 2.1 Milling tool for routing carbon-fibre-reinforced plastics, by Kendu®, made of submicrograin tungsten carbide (*top*), and with TiAlN coating applied by Metal Estalki® (*bottom*)

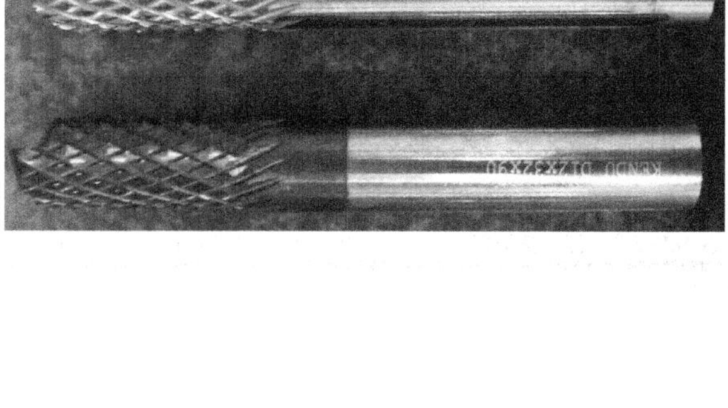

Secondly, there are small and medium companies that start from calibrated material rods, supplied by some of the former companies, and give form and geometry to cutting tools. This is the case of integral endmills, drilling tools and tailor-made tools. The natural markets for these companies are either very specific niches or special tailor-made tools built with user requirements.

2.1.1 High-speed Steel

This group of high-alloyed steels was developed at the early years of the 20th century. Basically they are high-content carbon steels with a high proportion of alloy elements such as tungsten, molybdenum, chromium, vanadium and cobalt. The mean hardness is 75 HRC.

The T series includes tungsten, the M series molybdenum, whereas vanadium produces the hardest of the carbides giving rise to the super-high-speed steels. The maximum working temperature of HSS is about 500 °C. Currently, HSS produced by powder metallurgy (HSS-PM) offers a higher content of alloy elements and a combination of unique properties: higher toughness, higher wear resistance, higher hardness and higher hot hardness. In Figure 2.2 a comparison of tool materials regarding hardness and bend strength is shown, in which the latter, directly related to toughness, is the main advantage of this type of tool material.

HSS and HSS-PM are excellent substrates for all coatings such as TiN, TiAlN, TiCN, solid lubricant coatings and multilayer coatings.

HSS-PM has many advantages in high-performance applications such as rouge milling, gear-cutting tools and broaching, and also in cases of difficult tapping, drilling and reaming operations. HSS-PM is used too in disc and bandsaws, knives, cold-work tooling, rolls, *etc*. However, for machining of tempered steels and very difficult-to-cut alloys HSS is not the first choice; tungsten carbide is a more recommended tool material (see Section 2.1.2).

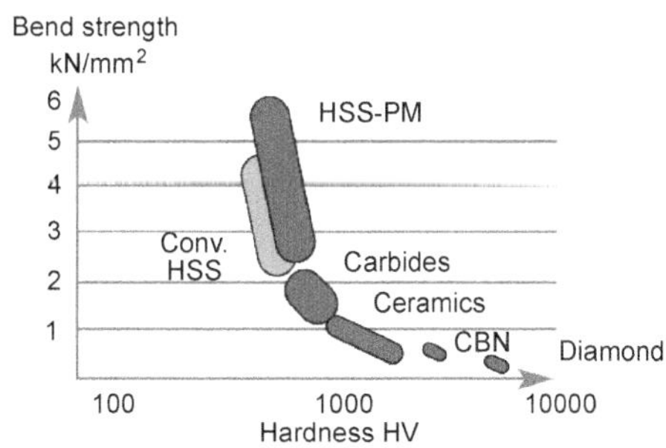

Figure 2.2 Bend strength *versus* hardness for tool materials (HSS Forum [1])

2.1.2 *Sintered Carbide (Hardmetal)*

Sintered carbide tools, also known as *hardmetal* tools or *cemented carbide* tools are made by a mixture of tungsten carbide micrograins with cobalt at high temperature and pressure. Tantalum, titanium or vanadium carbides can be also mixed in small proportions.

Therefore two main description factors define a hardmetal grade:

- The ratio of tungsten carbide and cobalt. The latter usually ranges from 6 to 12 % and it acts as binder. Cobalt has a high melting point (1493 °C) and forms a soluble phase with tungsten carbide grains at 1275 °C which helps to reduce porosity.
- The grain size, thus micrograin grades include particles smaller than 1 μm, and submicrograin are smaller than a half of a micron; the smaller the grain, the harder the hardmetal. Hardness increases with the reduction in binder content and tungsten carbide grain size, and *vice versa*, with values from 600 to 2100 HV.

Hardmetal tools are manufactured in two forms:

- Integral tools: they are manufactured by grinding a raw hardmetal rod, obtaining an endmill, a ball-endmill (Figure 2.3) or a drilling tool. The main advantage is the perfect balance of these rotary tools, but the main disadvantage is their high price, taking into account that only a little and very specific zone of the tool is worn by the cutting process. Several resharping of each tool are possible.
- Inserts: small pads with special geometry made with hardmetal, but they are fixed on toolholders made of steel. Turning tools and big milling discs use this configuration, which implies a rapid substitution of worn inserts.

Hardmetal grades are classified under the standard ISO 513 [2] into six groups, M, P, K, N, S and H, following a numerical scale for each of them. On the other hand, in the USA the C-x scale is used instead. The original concept of both classifications was to rate tungsten carbides according to the job that they had to do, and this led to a little clear scale in which no cobalt binder amount or grain size is specified. As consequence, tungsten carbide from different manufacturers may

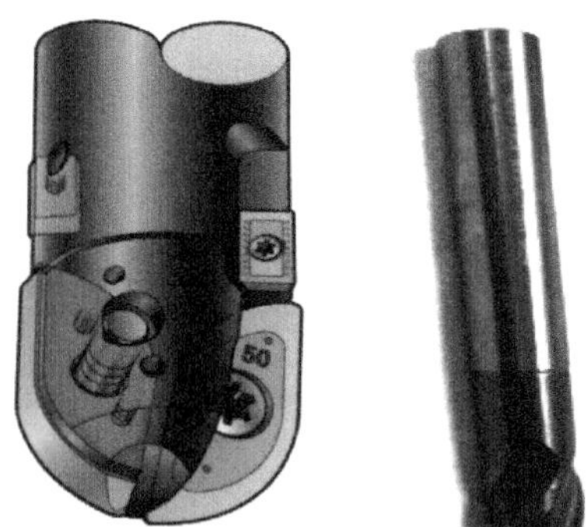

(a) (b)

Figure 2.3 Ball-endmill with inserts (a), and integral bull-nose endmill (b)

have identical designation but may vary considerably in performance. In Table 2.1 hardmetal grades offered by the company Ceratizit® are shown.

The ISO group recommendations are:

- P, indicated for low- and medium-carbon steels, and light alloyed steels;
- M, composed of sintered carbides, suitable for the stainless steels machining;
- K, oriented to cast irons and alloyed steels, and harder than the P and M series;
- H, for tempered and hardened steels;
- S, for heat-resistant alloys and titanium alloys;
- N, for aluminium alloys.

The two-digit number after the letter, from 01 to 40 (50 in P group) defines the hardness and toughness of the grade. The lower numbers correspond to the harder grades, whereas the higher are the tougher of them. K10 to K30 are the most used today.

Regarding the American classification, C-1 to C-4 are general grades for cast iron, non-ferrous and non-metallic materials, C-5 to C-8 are suitable for steel and

Table 2.1 Ceratizit® hardmetal grades

Grade	ISO code	Code USA	Grain size	TiC Ta(Nb)C	Binder	Density	Hardness			Transverse rupture strength TRS		K_{IC}
					%	g/cm³	HV 10	HV 30	HRA	MPa	P.S.J.	MPa.m$^{1/2}$
TUNGSTEN CARBIDE COBALT GRADE												
TSF22	K10–K20	C-2	Ultrafine	–	8.2	14.55	1970	1930	93.7	4400	638,000	7.5
TSF44	K10–K30	C-2	Ultrafine	–	12.0	14.10	1760	1730	92.7	4600	667,000	7.8
MG 12	K05–K10	C-3	Submicron	–	6.0	14.80	1820	1790	93.0	3500	507,500	8.2
TSM20	K10–K30	C-3	Submicron	–	7.5	14.75	1750	1720	92.6	3500	507,500	8.6
TSM33	K20–K40	C-2	Submicron	–	10.0	14.50	1610	1590	91.9	3700	536,500	9.4
MG 18						14.45	1680	1660	92.3	3700	536,500	9.4
CTS18D	K20–K40	C-2	Submicron	–	9.0	14.55	1610	1590	91.9	3600	522,000	10.4
CTF12A	K15	C-2	Fine	–	6.0	15.00	1650	1630	92.1	2600	377,000	10.2
HC10	K10	C-3	Fine	–	5.6	14.95	1760	1730	92.7	2150	311,900	9.2
H20X	K15	C2	Fine	–	6.0	14.95	1670	1650	92.2	2200	333,500	9.9
WC-TiC/TaNbC – COBALT GRADE												
S4X7	P30–P35	C-5	Fine	12.0	11.0	14.95	1490	1470	91.0	2300	333,500	11.6
CERMET												
TCN54	HAT–P20	–	–	–	14.1		1650	1630	92.1	2000	290,000	8.5
SILICON NITRIDE GRADE												
SNC 1	CN–K20	–	–	–	9.0		1550	1530	91.5	1100	159,500	6.5

steel alloys because these grades resist pitting and deformation, C-9 to C-11 are indicated for high-wear applications, and C-12 to C-14 are for impact cases. A common misconception is that higher grades have less cobalt binder and therefore are harder and fragile, but that is not true. For this reason and others the ISO standard is currently increasing in use.

A tool material derived from hardmetal is the cermet (ceramic–metal) type, sintered tungsten carbide also including TiC (carbide with hardness 3200 HV) and in some cases TiCN, but they typically have a nickel–chrome binder. New grades with TaNbC and MoC increase the tool-edge strength against the cyclic impacts typical of milling.

Tungsten carbide is very stable regarding chemical and thermal aspects of machining, and is very hard as well. In most cases, cemented carbide degradation starts from the cobalt binder and the tungsten carbide–cobalt cohesion.

2.1.3 Ceramics

Ceramics are very hard and refractory materials, withstanding more than 1500 °C without chemical decomposition. These features recommend them to be used for the machining of metals at high cutting speeds and in dry machining conditions. Unfortunately they are fragile, and ceramics without any reinforcement are only indicated for turning of continuous shapes. In milling the continuous impact at each tooth entrance in the machined part implies a high risk of chipping and tool failure.

Ceramic materials are moulded from ceramic powders at pressures more than 25 MPa, to be later on sintered at approximately 1700 °C.

Ceramic tools are based primarily on alumina (Al_2O_3), silicon nitride (Si_3N_4) and sialon (a combination of Si, Al, O and N).

Alumina tools can contain additions of titanium, magnesium, chromium or zirconium oxides distributed homogeneously into the alumina matrix to improve toughness.

(a) (b)

Figure 2.4 (a) Ceramic inserts: alumina (white), silicon nitride (grey), alumina with TiC (black), and (b) matrix of a reinforced ceramic (by Greenleaf®)

Silicon nitride ceramics present a higher resistance to thermal shock, with a higher toughness as well. These ceramics have a needle-like structure embedded in a grain boundary. This microstructure enhances fracture toughness. Their most typical application is the roughing of cast iron, even under unfavourable conditions such as heavily interrupted cuts. Silicon nitrides also are used to mill cast iron.

The ceramics reinforced by a non-homogeneous matrix of silicon carbide (SiC) whiskers ($Al_2O_3 + SiC_w$) are focused on the milling operation. Whiskers are fine-grained silicon carbide crystals similar to hairs. The whiskers form 20–40 % of the total ceramic, improving the tool toughness a lot, making them suitable for milling operations. Whisker-reinforced ceramics are successfully applied on hard ferrous materials and difficult-to-machine super alloys, especially in the case of the nickel-based alloy Inconel 718.

Ceramics are a very productive option in a lot of applications, but special care must be taken when machining is programmed. Tools must be kept hot throughout the operation (dry condition is the best) and shocks on tool edges at tool entrances and exits from the workpiece must be avoided. In turning, the *ramping* technique is highly recommended to reduce the notch wear in the cylindrical roughing of austenitic materials.

2.1.4 Extra-hard Materials

PCD and PCBN are extra-hard materials. There are several grades in the PCD and PCBN groups. As a rule of thumb, PCD is suitable for tools focused on machining abrasive non-ferrous metals, plastics and composites. Otherwise, PCBN finds applications in the machining of hardened tool steels and hard cast irons.

2.1.4.1 Diamond and Polycrystalline Diamond

PCD plates are obtained by a high temperature and pressure process where synthetic diamond grains are sintered with cobalt. Depending upon the machining operation, PCD is available in various grain sizes (Table 2.2). Thus, those grades with coarse grain sizes are used for making cutting tools with high wear resistance, but if very high surface finishing is required in the machined part, then ultra-micro grain sizes are preferred. Medium grain sizes are used for general-purpose cutting tools, since there is a balance between the high wear resistance of rough grain size and the good finish of ultra-micro grains.

Monocrystalline diamond (MCD) is natural diamond which enables the production of geometrically defined cutting edges with absolutely notch-free flutes. Natural diamonds often contain nitrogen which can produce varying hardness and thermal conductivity. This very expensive material is suitable for achieving very high surface finishes for mirror-bright surfaces, machining of non-ferrous materials, micromachining, dressing grinding wheels and machining of super alloys

Table 2.2 Some of the PCD grades by Sumitomo®

	Grade		
	DA90	DA150	DA200
Average diamond crystal size (μm)	50	5	0.5
Vickers hardness	10,000–12,000	10,000–12,000	8,000–10,000
Transverse rupture strength (kg/mm^2)	110	200	220
Product description	Coarse-grain diamond • Ultra-high abrasion resistance	Fine-grain diamond • High abrasion resistance • Excellent tool-edge sharpness	Ultra-fine-grain diamond • Superior tool-edge sharpness and toughness
Machining applications	• High-silicon aluminium • Graphite • Aluminium/grey iron bimetal • Ceramics • Tungsten carbide • Kevlar	• Low- and medium-silicon aluminium • Copper • Fibreglass • Carbon • Wood – plywood, fibreboard and hardwoods	• Plastics • Wood • Aluminium and copper applications where: – Low microfinish is required – Workpiece has severely interrupted surface

without burrs. Currently, the development of synthetic MCD in triangles and rectangles with an edge length of approximately 6–10 mm makes economically possible the use of this material for high-end applications.

2.1.4.2 Polycrystalline Cubic Boron Nitride

CBN is a polymorph boron-nitride-based material. Its high mechanical properties are due to its crystalline structure and its covalent link. It has been industrially produced since 1957, starting from hexagonal boron nitride put under high pressures (8 GPa) and temperatures (1500 °C). With a lower hardness (<4500 HV) than diamond (>9000 HV), CBN is the second-hardest synthetic material.

The CBN grains are sintered together with a binder to form a composite, PCBN. The size, shape and ratio of CBN/binder define the different PCBN grades (Table 2.3, Figure 2.5). The content of CBN crystals ranges from 40 to 95 %, whereas binder may be Co, W or ceramic. If interrupted cutting (milling) of iron castings is to be performed, high CBN content and Co matrix grade is recommended. Low CBN content and ceramic matrix can be used in finishing operations. PCBN is typically recommended in the turning, milling and drilling of pear-

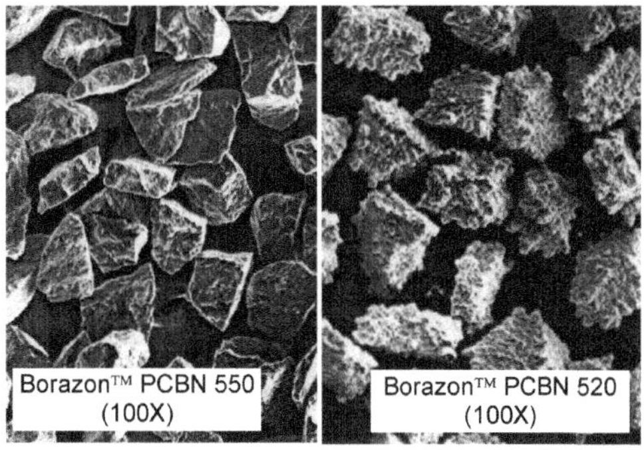

Figure 2.5 Two grades of PCBN, Borazon™ by General Electric®

litic iron castings, both grey and ductile, but should not be used for ferritic iron castings. Ferrite is highly reactive, and produces the degradation of the CBN because of the diffusion of boron within the ferritic matrix.

Degradation of PCBN in the turning of thermal sprayed layers on turbine axles due to a complex chemical attack of coolant has also been reported [3].

Using PCBN, cutting conditions can be very high, for example $V_c = 1800$ m/min, $f_z = 0.31$ mm/z in the machining of lamellar grey iron casting (GG25) with tool life of 1200 m/tooth. In the case of martensitic (55 HRC) or white (55 HRC) iron castings, cutting speed within the range 100–200 m/min using feed rates of about 0.15 mm/z can be used [4]. Recommendations given by tool manufacturers are more conservative, with values of 300–900 in the case of grey irons and 100–300 m/min for ductile irons.

If cutting speeds over 1500 m/min are used, an aluminium oxide layer may appear on the cutting edge, resulting from the adhesion of aluminium inclusions from the iron casting. At lower cutting speeds that layer protects the edge from wear. However, over 1500 m/min the temperature at the edge becomes too high (around 1400 °C), which collaborates with a complex chemical reaction between the PCBN binder material and the silicon from the casting, giving rise to TiB2 and TiCN products.

As far as the tool geometry is concerned, in the case of endmills the best results have been obtained when a small helix angle was used (about 2°). Supposedly, the cutting is more stable when using low helix angles. In [5], endmills with helix angles of 2° and 30° are compared, both under severe machining conditions: $V_c = 1500$ m/min, $f_z = 0.02$ mm, $a_p = 18$ mm, $a_e = 0.05$ mm, tool diameter = 60 mm and no coolant. Before the end of the tool life was reached, the former could machine a length of 1800 m, whereas the length machined by the latter was 850 m.

PCBN commercial tools consist of a PCBN layer placed on a hardmetal tool body. Again, the tool/toolholder balance is a crucial factor when choosing integral rotary tools.

Usually, PCBN tools present a very simple geometry. However, the CBN300 chipbreaker (by SECO®) is designed with an increased rake angle. This leads to lower cutting forces and lower levels of transferred energy, resulting in a lower temperature levels in the cutting zone.

Table 2.3 Some of the PCBN grades by Sumitomo®

	Grade						
	BN100	BN250	BN300	BN500	BN600	BNX20 (US300)	BNX10
CBN content (%)	85	60	60	65	90	60	50
CBN crystal size (μm)	3	1	0.5	4	2	3	4
Primary binder material	Titanium nitride	Titanium nitride	Titanium nitride	Titanium carbide	Co–Al	Titanium nitride	Titanium nitride
Vickers hardness (HV)	3900–4200	3200–3400	3300–3500	3800–3500	3900–4200	3200–3400	2800–3000
TRS (kg/mm^2)	85	105	115	105	105	105	85
Recommended for machining	• Grey cast iron • Powdered metal	• Light- & medium-interrupted hardened steel	• Severely interrupted hardened steel	• Nodular iron • Grey cast iron • Alloyed iron	• Grey cast iron • Powdered metal • Chilled cast iron • Ni/Co-based superalloys • Ni-hard iron	• High-speed continuous turning of hardened steel	• High-speed continuous hardened steel finishing

The use of PCBN depends of several factors to be taken into account. Thus, after several milling tests on hardened steels for mould making, these factors were gathered and evaluated (Table 2.4).

PCBN suits fine the turning of continuous surfaces, for example the turning of brake discs. In ball-endmilling the major problem to be solved is that cutting speed at the tool tip is zero, and as a result, the PCBN suffers high mechanical stresses. The best method to overcome this problem is the use of 3+2-axis machines, placing the tool at angles of 15–20° with respect to the surface to be machined.

A finishing test on hardened steels of a small mould of 85 mm side length is here selected as example. Before the introduction of PCBN and high-speed milling this piece was manufactured by electrodischarge machining because of its small radii and deep cavities. The applied parameters with PCBN were axial depth of cut $a_p = 0.2$ mm, radial depth of cut $a_e = 0.2$ mm, cutting speed $V_c = 1000$ m/min, feed per tooth $f_z = 0.15$ m/tooth and spindle speed $n = 24{,}000$ rev/min, producing the mould in less than 17 min. Lead time for this part was reduced by more than 500 %.

Table 2.4 Success factors for PCBN application of high-speed milling (HSM) on iron castings and tempered steels

Factors		Probability of success of HSM on dies and moulds with PCBN tools		
		High	Medium	Low
Material		Free ferrite lower than 5 %	→	Free ferrite higher than 10 %
Machine	Spindle speed	High-speed machine (<15,000 rpm)	→	Machine with conventional speeds
	Number of axes	5 axes	→ 3 + 2 axes	→ 3 axes
Cutting fluid		Dry	→ Oil mist under high pressure	Coolant
Process	Trajectory	Continued	→	Interrupted
	Strategy	Inclination of the tool towards the feed direction and climb milling	→	Cutting with the top of the tool ($V_c = 0$)
	Computer-aided dsign/manufacture	CNC optimized	→	CNC without optimization
	Simulation and virtual optimizing	Control of the feed rate for a constant chip volume	→	Stepped feed motion
Geometry		Big surfaces of low complexity	→	Small surfaces with high complexity and cavities

2.2 Coatings

A tool coating is a layer with thickness ranging from 2 to 15 μm solidly deposited and bonded to the tool substrate to improve the cutting-tool performance (see Figure 2.6), and applied after the tool is shaped. Coatings provide a hard, chemically stable surface and thermal protection to tools, improving their performance during cutting.

2.2.1 Historical Introduction to Physical Vapour Deposition Coatings

Physical vapour deposition (PVD) coatings are ceramic materials usually applied in 1–15 μm thicknesses on tools made of steel and hardmetals. They were developed industrially in the 1970s to provide the ability of ceramics to withstand high

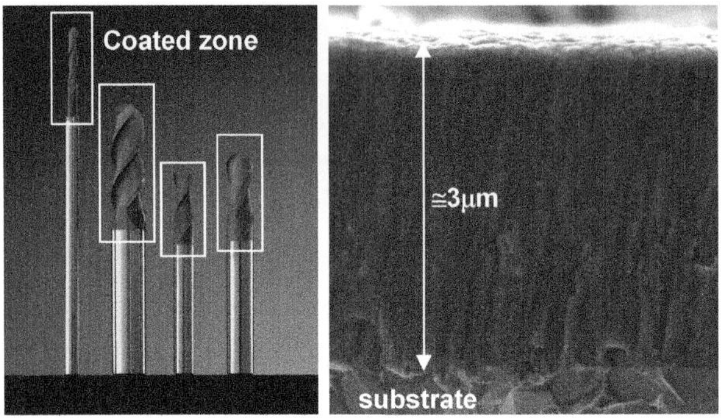

(a) (b)

Figure 2.6 (a) PVD-coated hardmetal tools, and (b) cross-section of an AlTiN coating. Source: Oerlikon®

temperatures to substrates much tougher than ceramics such as HSS and hardmetals. This combination resulted in one of the most successful developments in the last 30 years in cutting-tool materials and since then a great improvement in cutting speeds and productivity has been achieved.

Chemical vapour deposition (CVD) coatings were already commercialized for carbide inserts in previous years but those based on PVD technology were the ones which resulted in the broadest market impact. This success was due to the possibility to process PVD coatings at much lower temperatures than CVD coatings, 400–500 °C against 900–1000 °C, which enabled the use of PVD coatings for HSS tools. But there was also a great difference which helped promote the use of PVD coatings: the ability to control thicknesses on the edges accurately. This latter property guaranteed a sharper coated edge coated with PVD compared to an edge coated by CVD. There were also other properties such as higher intrinsic hardness and compressive stresses which helped promote their use against CVD coatings; this latter property favours the inhibition of crack growth in tool edges which are exposed to impact. The freedom to coat by PVD without chemical interaction with the substrate was also a great advantage, contrary to CVD coatings which easily interact with the substrates, occasionally producing brittle carbides at the interfaces. Lastly, the ease of recoating and resharpening PVD-coated tools, against CVD-coated tools, opened a large industrial market highly sensitive to cost-reducing opportunities.

2.2.2 Industrial Evolution of Different Compositions

The first commercial coating was a titanium nitride, and since then most of the industrial coatings have been based on nitrides. It was 1979 when Oerlikon® (previouly Balzers) began the production of TiN coatings based on electron beam ion-

Table 2.5 Current tool coatings, Platit®

Coating	Colour	Nanohardness (GPa)	Thickness (μm)	Friction (fretting) coefficient	Max. usage temperature (°C)
TiN	Gold	24	1–7	0.55	600
TiCN-MP	Red-copper	32	1–4	0.20	400
TiCN	Blue-grey	37	1–4	0.20	400
CrN	Metal-silver	18	1–7	0.30	700
CBC	Grey	20	0.5	0.15	400
AlTiN	Black	38	1–4	0.70	900
μAlTiN	Black	38	1–4	0.30	900
TiAlCN	Burgundy-violet	33	1–4	0.30	500
Cromvic	Grey	20	1–10	0.15	400
Gradvic	Grey	20/33	1.5–5	0.15	400
cVic	Grey	20/37	1–5	0.15	400
ZrN	White-gold	20	1–4	0.40	550
AlCrN	Blue-grey	32	1–4	0.60	900
nACo	Violet-blue	45	1–4	0.45	1200
nACRo	Blue-grey	40	1–7	0.35	1100
nATCRo	Blue-grey	42	1–4	0.40	1150
nACo3	Violet-blue	45/34	1–5	0.45	1200/900
nACRo	Blue-grey	40/34	1–5	0.35	1100/900
nATCRo	Blue-grey	42/34	1–5	0.40	1150/900

plating technology and this conspicuous golden coating was to play the leading part in making PVD coatings very popular.

The next generation of industrial coatings was composed of chromium nitride (CrN) and titanium carbonitride (TiCN), the first of them focused on forming tools and cutting soft metals, broadening the application of PVD coatings. The other one focused on enhancing the hardness of TiN coatings from 2300 HV to 3200 HV, which resulted in an overall improvement of the performance of materials usually coated with TiN.

But it was not until the late 1990s when a major change arrived in coating technology with the production of TiAlN coatings. The addition of aluminium to the TiN-based composition provided not only a higher hardness such as 3300 HV but a remarkable improvement which was enhanced high-temperature behaviour. In order to explain the latter property it must be noted that during the use of a coating in a cutting process the edge must withstand temperatures of several hundred de grees Celsius. With both TiN and TiCN there is an unavoidable hardness reduction above 500 °C, therefore limiting their use in high-speed or dry conditions which result in higher temperatures at the cutting edge.

The effect of the aluminium alloying resulted not only in a greater hardness at temperatures of up to 900 °C, but also it provided a much better oxidation resistance up to that temperature. Both properties, hardness at high temperatures and oxidation resistance up to 900 °C, opened a new field of cutting conditions for the most advanced tools, meaning higher cutting speeds and dry cutting with lifetimes comparable to cutting tools coated with TiN working at moderate conditions with lubrication. This was a great leap for cost saving in advanced manufacturing processes.

These coatings were commercialized by most of the leading coating technology manufacturers: Platit® with Universal TiAlN, Oerlikon® with Balinit Futura and Cemecon® with Tinalox.

On the other hand, several new coatings were developed for different cutting materials and tool types, and also for forming tools. Table 2.5 summarizes some of the coatings offered by Platit's technology.

2.2.3 Current Trends in Coatings for Hard Machining

The next stage in the evolution of TiAlN coatings came with those usually known as AlTiN coatings, for their higher aluminium content. As noted by coating developers, higher aluminium content implies a better thermal resistance. The reason for that behaviour was the nanostructuring of the coating into TiAlN crystallites in a cubic AlN-based matrix. This nanostructure, which compared to the microcrystalline TiAlN was more stable at high temperatures, enabled a further increase in lifetime of carbide endmills for high-speed cutting. The most remarkable example of these successful coatings was their use on ball-nose endmills for cutting hard tempered steels such as those employed for moulds made of tempered steels. The coatings under this composition are Platit® AlTiN, with up to 67% aluminium, Oerlikon® Xceed and Cemecon® Hyperlox.

However, a new trend in high-temperature nanostructure control was set when Hitachi® unveiled endmills coated with TiAlN–TiSiN coatings and soon after, Platit® did so with AlTiSiN coatings with the nACo™ trademark. Figure 2.7 shows the wear behaviour of carbide ball-nose endmills coated with different

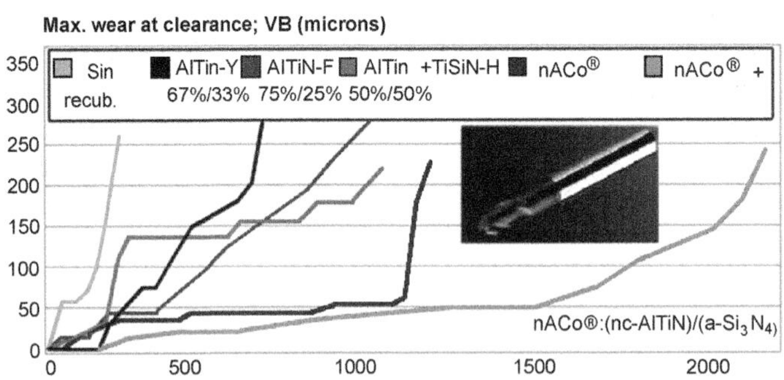

Figure 2.7 Results of wear measurements for solid carbide endmills ($z=2$, $\varnothing=10$ mm, rpm = 18500, $f_z=0.18$ mm, $a_p=0.25$ mm, $a_e=0.6$ mm, minimum quantity of lubricant). Workpiece: 1.2343 tool steel (57 HRC). Source: Platit

coatings. The step increase in wear resistance from an AlTiN coating to a nACo™ AlTiSiN coating is remarkable. Also shown is the result for higher Si-content nACo™-coated endmills.

Silicon-containing coatings have been adopted by tool manufacturers and end users for improving more hard-machining conditions. The success of employing silicon alloying ensures that a fine nanostructure is maintained up to 1200 °C, therefore, the hardness loss at high temperature is minimized thanks to silicon in the coating, which surrounds TiAlN crystallites as a silicon nitride binder.

Another machining process requiring high-temperature hardness is titanium milling. It is well known that great heat is involved in the cutting operation of titanium. In this case, as shown in Figure 2.8, the addition of silicon in the nACo™ coating produces the best result.

The year after introduction of AlTiSiN coating to the market a new coating was presented by Oerlikon: Balinit Alcrona. This coating is an AlCrN coating intended for expanding the capabilities of TiAlN coatings especially where high oxidation resistance is required. The hardness of AlCrN coating is similar to that of TiAlN, but what makes this coating outstanding is its high oxidation resistance, up to 1200 °C. That is thanks to the growth of a stable $(Al,Cr)_2O_3$ oxide during cutting instead of the $TiO_2 + Al_2O_3$ oxides which grow in TiAlN coatings.

However, the main achievement of AlCrN coatings is limited to a lifetime extension of hardmetal tools under standard cutting conditions and its successful use for hardmetals is usually far from hard-machining conditions. In order to overcome this limitation Platit developed a new silicon-containing AlCrN coating: nACRo™. This last AlCrSiN-based coating has been successfully applied in hobbing, drilling and milling when both high temperature resistance and oxidation resistance of the coating are required.

Current trends in coating technology for hard machining can hardly be explained on a general basis as the coating applications are becoming more specialized than ever; for similar machining methods different approaches are found.

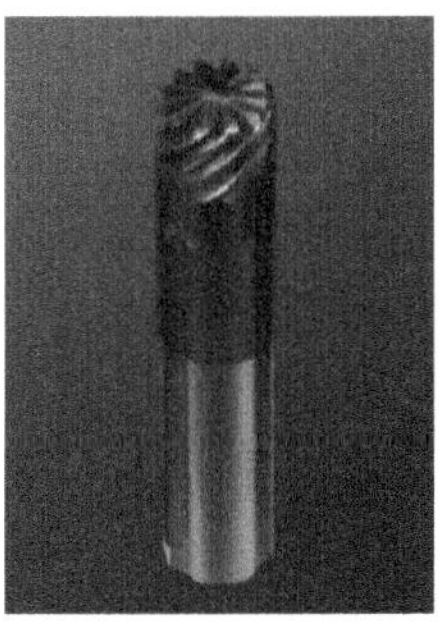

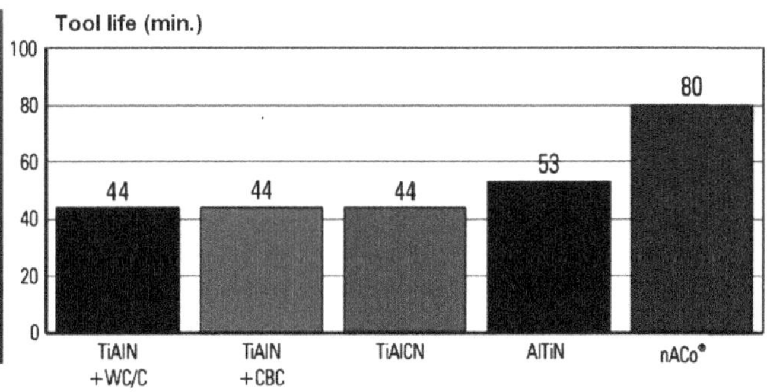

(a) (b)

Figure 2.8 (a) Carbide mill, and (b) tool life for carbide mills ($z = 12$, bull-nose radius: 1.2–1.9 mm, $\varnothing = 20$ mm, $V_c = 250$ m/min, $f_z = 0.11$ mm, $a_p = 0.5$ mm, $a_e = 1.1$ mm). Workpiece: TiAl6V4. Source: Platit

From the material point of view, alloying of TiAlN coatings with different alloying elements opens endless possibilities: TiAlCrN, TiAlCrSiN and TiAlCrYSiN compositions are reported by several researchers and even addition of Zr, V, B or O to coating compositions.

2.2.4 Coating Selection and Optimization for Hard Machining

One parameter for a fixed coating design is a proper selection of the thickness in order to provide a life long enough to the edge but avoiding the adhesion failure of the coating due to internal compressive stresses.

The selection of the proper coating structure involves combining the best properties of the following structures:

- Monobloc coating (monolayer of the same composition): used when there is no impact or when cutting forces are low.
- Bilayer coating for combining good properties of an inner layer near the substrate and upper layer; for example, when a hard coating is needed and top lubricant coating is needed for better chip flow.
- Multilayer coating to improve the shear strength of the coating, avoiding crack propagation between different layer materials.
- Adhesion layers: addition of a thin adhesion layer of 0.05–0.2 μm to increase the adhesion of the next layer.
- Triple coatings: a novel approach by Platit to optimize the coating structures, consisting of a good adhesion layer, a tough core layer and a hard and temperature-resistant top layer.

On the other hand, there is an even more important condition related to cutting-edge preparation before and after coating. One of the main obstacles to advanced coating success for hard-machining processes is the edge condition before coating. The more lifetime an advanced coating is able to provide the more sensitive it is to starting conditions in the edges. Therefore, along with the high-performance coating development a new approach has been required to stabilize the lifetime of the coated tool and new edge-finishing processes have been required. Figure 2.9 shows the great effect of the cutting edge radius on the lifetime of an endmill. As can be seen, there is a big difference between no radius and the optimum one.

But the coating surface can also be improved for better edge stability. Industrial PVD coatings are produced by arc technology, more economical and more suitable for providing stable quality to coatings. However, its main drawback is the presence of droplets in the coating surface, which originate from the target melting during the arc burning. These droplets are bonded to the coating surface and are responsible of most of the coating roughness. Coating roughness on the edge creates a deleterious effect on the lifetime stability, therefore droplet removal processes are usually performed for high-end tools. The effect of one of these processes is shown in Figure 2.10.

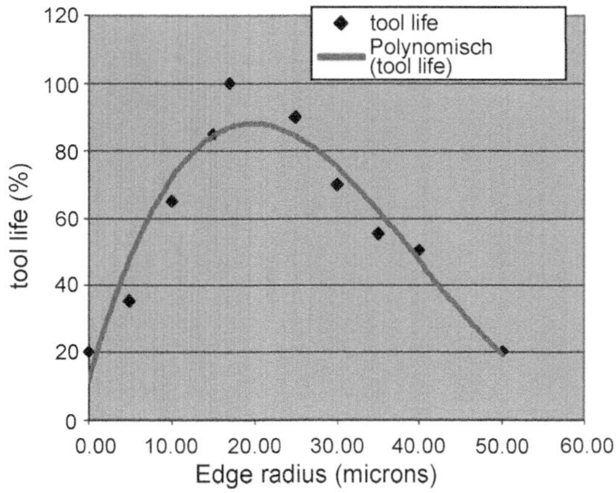

Figure 2.9 Tool life for carbide mills nACRo coated ($z = 4$, bull-nose radius 1.2–1.9 mm, $\varnothing = 10$ mm, $V_c = 150$ m/min, $f_z = 0.05$ mm/z, $a_p = 1.5 \times \varnothing$, $a_e = 0.25 \times \varnothing$). Workpiece: 1.2379. Source: Platit

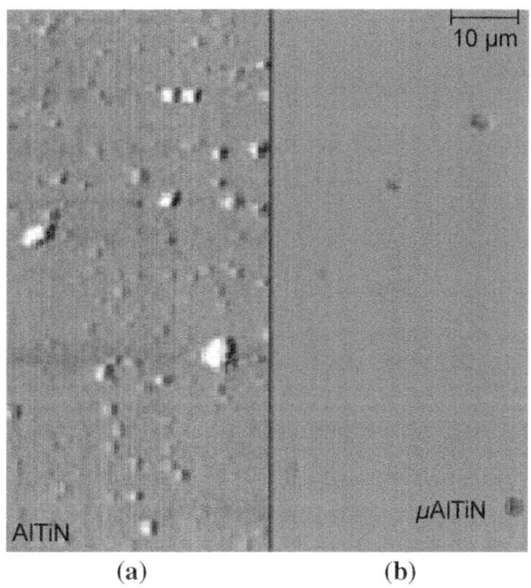

Figure 2.10 Images of AlTiN coatings (a) before and (b) after surface treatment for droplet removal, by Platit®

2.3 Tool Wear

Tool wear is caused by the continuous action of the chip removal process, and can be located in two tool zones:

- wear on the rake face, which usually gives rise to a crater-like pattern;
- wear on the flank or clearance face, due to the high friction of tool edge with the fresh machined surface. It looks like a typical abrasion pattern.

All tool wear types are described in the corresponding ISO standards. In Figure 2.11 a turning tool is shown, where the main wear zones and the way to define them is based on ISO 3685, *Tool-life testing with single-point turning tools* [6].

Figure 2.11 Wear on a turning tool, ISO 3685

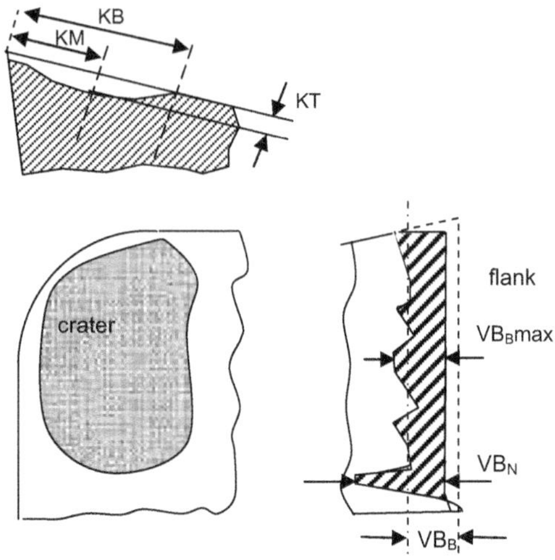

2.3.1 Tool Wear in Turning

Turning is a continuous operation with constant cutting force. However, tools undergo constant heating derived from the shear deformation energy and friction, which cause a high temperature at the tool/chip interface. The high temperature at the tool rake face is a principal wear factor in turning, being for austenitic steels, superalloys or titanium alloys even more than 600 °C.

Basically four wear mechanisms are possible in turning:

- *Crater wear:* a chemical/metallurgical wear due to de diffusion and adhesion of small particles of the tool rake surface on the fresh chip. A mechanical friction also collaborates in causing a scar-like shape on the rake face which usually is parallel to the major cutting edge. Crater wear is frequent in the turning of titanium alloys (see Figure 2.12) and other low thermal conductivity materials.
- *Notch wear:* a combination of flank and rake face wear which occurs just in the point where the major cutting edge intersects the work surface (it coincides with the depth of cut line). It is very typical in the turning of materials with

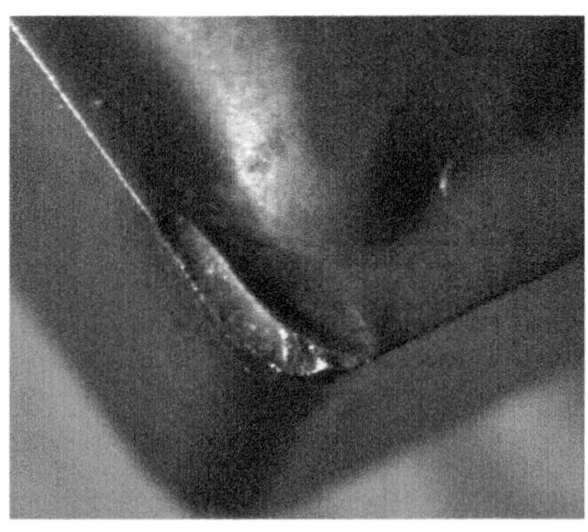

Figure 2.12 Crater wear in the turning of Ti6Al4 V, very close to the tool edge

tendency to surface hardening due to mechanical loads. Thus, previous tool passes rub the fresh machined surface increasing the hardness of the outer layer (this hardened skin is only few microns thick). Notch wear is common in the turning of austenitic stainless steels and nickel-based alloys.

- *Flank wear:* this type of wear is placed on the flank (relief) face (see Figure 2.13). Wear land formation is not always uniform along the major and minor cutting edges of the tool. It is the more common in the case of hard materials where no chemical affinity between tool and material exists, abrasion being the main wear mechanism.
- *Adhesion:* due to the high pressure and temperature, welding occurs between the fresh surface of the chip and tool rake face. This is a considerable welding if materials have metallurgical affinity and causes a thick adhesion layer, and a posterior tearing of the softer rubbing surface at high wear rate. Adhesion is usual in the case of aluminium alloys in dry or near-to-dry conditions, but it is not common in hard machining.

In most machining processes, flank wear is the type to control because it implies a significant variation of tool dimensions and therefore in the dimension of machined parts. Values of 0.3–0.5 mm are the maximum accepted, the former value for finishing and the latter for roughing.

Figure 2.14 illustrates a typical evolution of mean flank wear (VB_B) along time, for different cutting speeds. As occurs in all friction cases, the relative speed between the two contact surfaces is the leading factor of degradation. The *wear curve* is divided into three stages, similar to the friction wear of other mechanical components:

- The zone AB where the sharp new edge is worn rapidly. The initial wear size is $VB = 0.05–0.1$ mm.
- The zone BC, where wear rate is constant and slowly increases. This zone starts from 0.05 to 0.6 mm onwards.
- The zone CD, where wear ratio is very high. When this zone is reached a new tool must replace the worn one or resharpening must be performed before tool breakage.

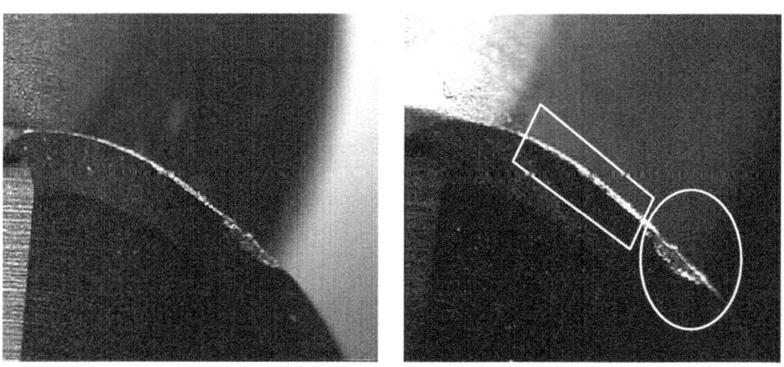

Figure 2.13 Flank at two times in ball-end milling in the finishing of a mould on 50 HRC steel. In the rectangle is the mean flank wear (*VB1*), and the circle indicates the maximum flank wear (*VB3*)

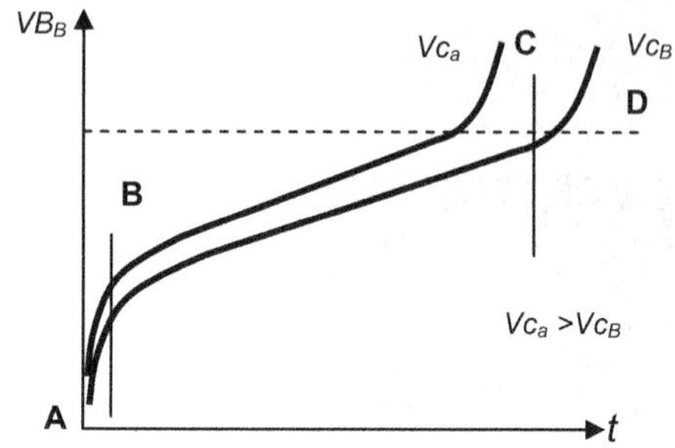

Figure 2.14 Flank wear evolution for different cutting speeds

2.3.2 *Tool Wear in Milling*

On general lines, those aspects commented upon above for turning are also valid for milling. The standard ISO 8688 [7] describes the main wear patterns and localizations, shown in Figure 2.15.

- *Flank wear (VB):* the loss of particles along the cutting edge, that is, in the intersection of the clearance and rake faces, being observed and measured on the clearance face of endmilling tools. Three different measurements are possible:
  - *Uniform flank wear* (*VB*1): the mean wear along the axial depth of cut.
  - *Non-uniform* flank wear (*VB*2): irregular wear in several zones of the cutting edge.
  - *Localized flank wear* (*VB*3): wear usually found in specific points. One type is that placed just in the depth of cut line, the notch wear (VB_N), typical of materials susceptible to mechanical hardening.

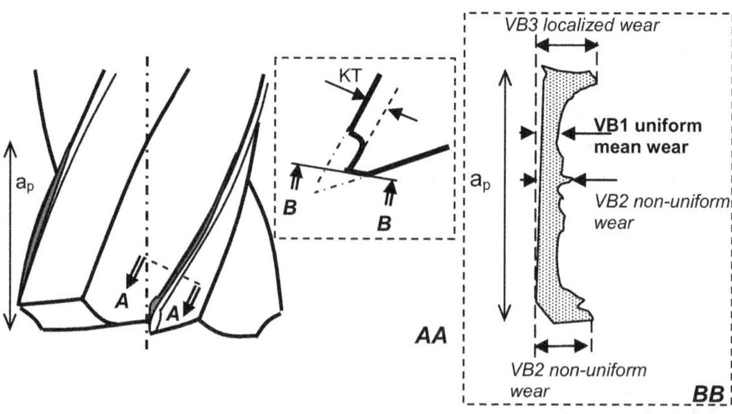

Figure 2.15 Wear of endmilling tools, from ISO 8688

- *Wear on the rake face (KT):* this is located on the internal flutes of endmills. The most typical is the crater wear (*KT*1), a progressive development of a crater oriented parallel to the major cutting edge.
- *Chipping (CH):* irregular flaking of the cutting edge, at random points (see Figures 2.16 and 2.17). It is very difficult to measure and prevent. It consists of small tool portions breaking away from the cutting edge due to the mechanical impact and transient thermal stresses due to cycled heating and cooling in interrupted machining operations.
- *Uniform chipping (CH1):* small edge breaks of approximately equal size along the cutting edge engaged on material.
- *Non-uniform chipping (CH2):* random chipping located at some points of the cutting edge, but with no consistency from one edge to another.
- *Flaking (FL):* loss of tool fragments, especially observed in the case of coated tools.
- *Catastrophic failure (CF):* rapid degradation of tool and breakage.

Mean flank wear size is the usual tool life criterion, due to it implying a significant variation of tool dimensions and therefore in the dimension of the machined part. Values of 0.3–0.5 mm are the maximum accepted, the former for finishing and the latter for roughing. Chipping greater than 0.5 mm is also a tool life criterion. In low machinability alloys several wear types appear simultaneously, adding and multiplying their negative effects [8] (see Figure 2.18).

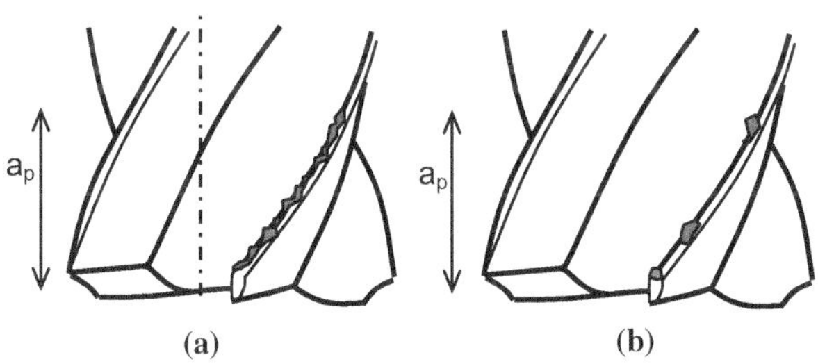

Figure 2.16 Chipping: (a) *CH*1, and (b) *CH*2

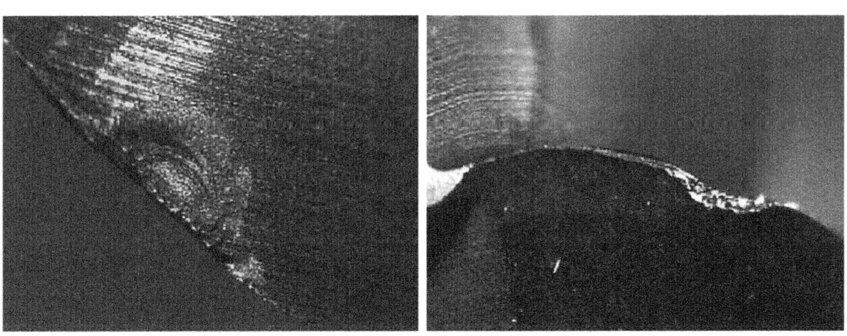

Figure 2.17 Chipping of a ball mill, after working on a tempered steel to 55 HRC

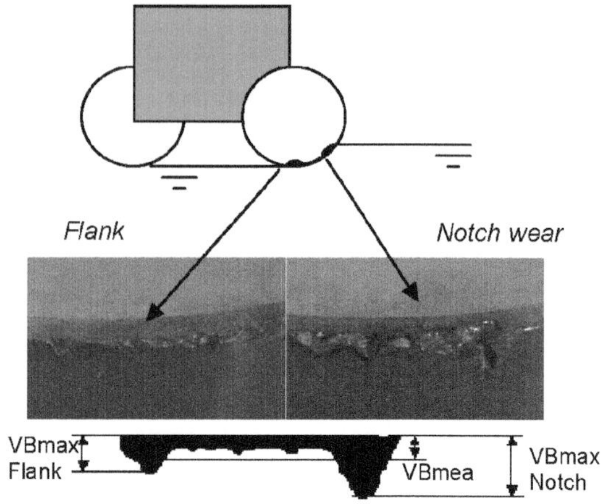

Figure 2.18 Two wear types when milling Inconel 718 with round-type whisker reinforced ceramics

2.3.3 *Tool Life*

Tool life is the time before a determined tool wear is reached. Since the first extensive experiments by Taylor in 1907, it has been known that cutting speed is the most influential parameter on tool life for a raw-material–tool couple. The so-called Taylor equation establishes that:

$$\frac{v_c}{v_r} = \left(\frac{T_r}{T}\right)^n \quad v_c T^n = v_r T_r^n, \tag{2.1}$$

where:

n is an experimental constant for each tool–material couple;
v_c is the cutting speed;
T is the tool life;
v_r is the reference speed at which a known tool life T_r is reached.

There are some variations including other machining parameters affecting tool life, for example:

$$V_c \cdot f_z^x \cdot a_r^y \cdot T_{VB}^n = C_{VB} \cdot VB^m, \tag{2.2}$$

where f_z is the feed per tooth, a_r is the radial width of cut, T_{VB} is the time to reach a determined VB, C_{VB} is a constant from experimental tests, and VB varies with the criteria used in the reference experiments. The x, y, m are characteristic of each tool-material couple.

Taylor parameters are usually known for common steels and free-machining materials, but difficult to find for low-machinability alloys. This difference derives form the fact that the final value of components usually made in common steels depends a lot on manufacturing costs; therefore the maximum use of each tool is a very important aspect to be economically competitive. However, components usually made of special alloys or of tempered steels are high-end products, and the final value of the component depends more on the machine cost per hour or the raw material itself. In this context the Taylor approach is not that interesting and thus few data about tool life appear in the literature.

Tool life is usually measured (a) in time, when constant machining parameters are used in a manufacturing process and the customer tries to compare similar tools from different suppliers, (b) in metal removal volume if roughing operations is being performed, or (c) in machined length if a finishing operation is considered. However, these three values are related by means of the machining parameters and process basic equations and they can be graphed in the same record (see Figure 2.19).

Some tool manufacturers make a special coating layer with a different colour on new inserts to make easy the measurement and detection of wear, such as that shown in Figure 2.20. The golden TiN coating applied to the inserts' clearance surfaces simplifies wear detection and thus avoids the unnecessary waste of unused cutting edges. The grey TiCN rake face minimizes negative tensile stress and improves adhesion and toughness.

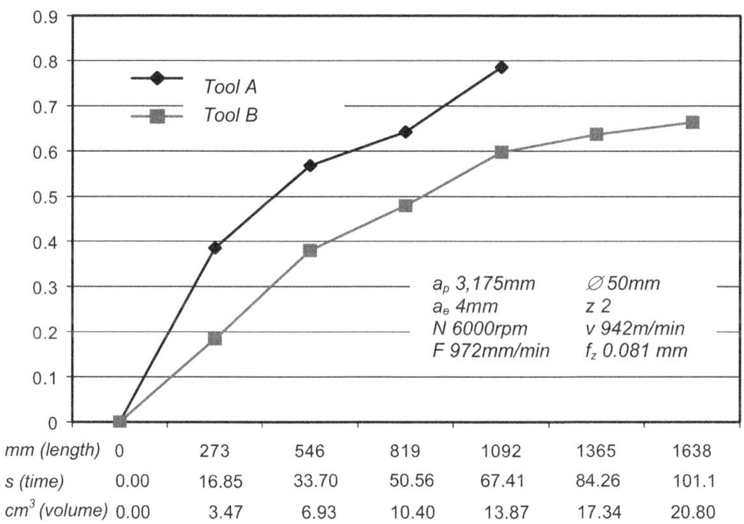

Figure 2.19 Typical tool life curves for two tools, flank wear *vs.* cut length, machining time and removed chip volume

Figure 2.20 Insert Tiger-Tec™ by Walter®

2.4 Cutting Fluids

Throughout a machining process, as much as 97 % of the mechanical energy is converted into thermal energy: 80 % of the heat is generated in the primary shear zone, 75 % of which is evacuated by the chip and 5 % goes to the machined part; 18 % of the total thermal energy is produced at the tool–chip interface, and 2 % comes from the tool-workpiece interface. These conditions of friction and temperature cause tool wear by different physical mechanisms explained in the previous section, giving in the result a poor surface finish and lack of precision. In Figure 2.21 the thermal field of a turning tool is shown, in the stationary thermal regime reached after several machining seconds.

Cutting fluids are used to reduce the negative effects of heat and friction on the tools and workpieces. The fluid produces three positive effects in cutting: (a) cooling, (b) lubrication between the chip and rake face of the tool, and (c) evacuation of chips towards the chip collecting system.

There are various types of cutting fluids, oils, oil–water emulsions, pastes, gels, mists and gases (liquid nitrogen and CO_2). They are obtained from petroleum distillates, plant oils or other raw ingredients.

For different reasons the reduction and even the total elimination of cutting fluids is advised. On the one hand, the cost of the life cycle of the cutting fluid (filtration, purification and elimination of residues) has a direct repercussion on the manufacture costs. On the other hand, the current environmental concern imposes heavy limitations on the use of hazardous substances (such as cutting fluids). Thus, in industrialized countries, strict regulations related to the use of cutting fluids are being developed. These regulations are increasingly restrictive with respect to the use of lubricants.

Because of the above-mentioned reasons dry machining would be of maximum interest, but this is somehow non-viable in aluminium and light alloys due to the tendency of these materials to adhere to the tool edges. Nor is it possible in the case of titanium, nickel or stainless steels due to the very high temperatures reached at the tool/chip interface. Therefore, taking into account the impossibility of dry machining, a technique involving minimal consumption of cutting oil called minimum quantity of lubricant (MQL) can be applied. This technique con-

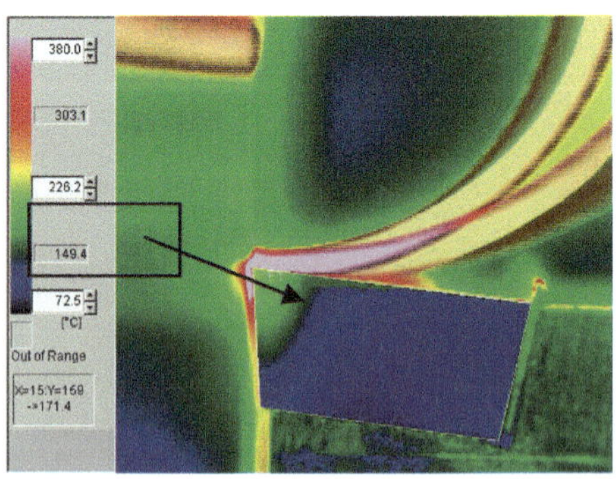

Figure 2.21 Infrared measurement of cutting temperatures, at $V_c = 137$ m/min and $f_z = 0.08$ mm, in a common steel turning

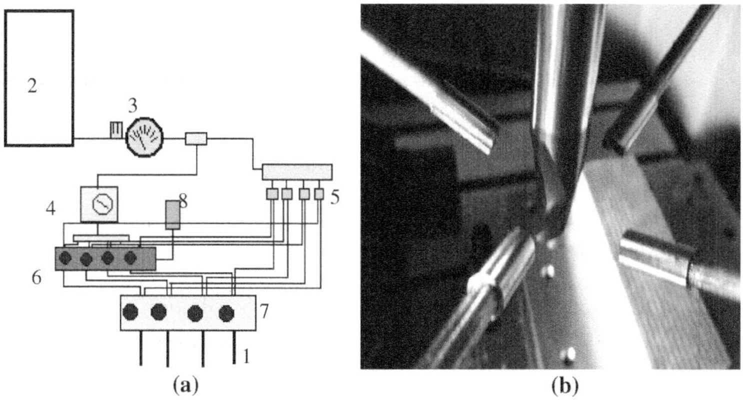

Figure 2.22 (a) Configuration of an MQL system, and (b) detail of the nozzles, four in this case

sists of the injection of a high-speed air jet with micro-drops of biodegradable oil in suspension.

A typical and modern MQL system is detailed in Figure 2.22. The system operates with pressurized air (10–12 bars). The pressurized air arrives to the system and it enters a maintenance unit (2); afterwards the air goes through a pressure regulator (3). Then, part of the air arrives to a subsystem where it produces the impulsion of the oil, regulated through a frequency meter (4) and several pumps (6) which provide the quantity of oil to be supplied by each nozzle at each instant. Oil is impelled up to the nozzle (1), where the mixture of oil with air is produced. The simultaneous effect of pressure and speed of the air in the exit nozzle sprays the oil. The obtained oil drops are below 2 μm diameter.

In the machining of hardened and tempered steels (more than 40 HRC in finishing conditions) dry or near to dry machining is a common option. In addition, the pressurized air injection is used to take away chips from the cutting zone. Cryogenic cooling by means of liquid nitrogen is now researched for titanium and nickel alloys [9]. However when difficult-to-cut alloys are machined the most common technique is the emulsion coolant, 5–10 % oil in water.

2.5 Tool Geometry

A cutting tool presents a main cutting edge and several faces. Many tools have another secondary cutting edge (minor edge). The shape of edges and angles between faces influence the machining performance greatly. In Figure 2.23 the basic geometry for a single-point cutting tool (that is, a turning tool) and for a multiple-point cutting tool (an endmilling tool in this case) are shown.

The definition of tool geometry is explained in the ISO 3002/1 [10]. Here two reference systems are described: *tool-in-hand* and *tool-in-use*.

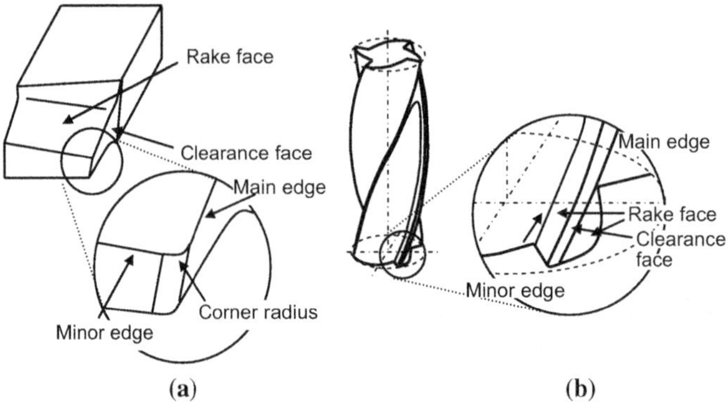

Figure 2.23 Basic geometry for (a) turning and (b) endmilling tools

In the tool-in-hand reference system three planes are defined:

- P_r: tool reference plane, parallel to the tool base or contains the axis of the rotational tool;
- P_f: assumed working plane, perpendicular to P_r and contains the feed direction;
- P_p: tool back plane, perpendicular to P_r and P_f;
- P_n: edge normal plane, perpendicular to the edge in each point.

In Figure 2.24, P_r, P_f and P_p for a turning and a milling tool are shown. Using these planes several angles are measured:

- κ_r: position edge angle (measured in P_r);
- κ'_r: position edge angle of the minor edge (measured in P_r);
- γ_n: normal rake angle (measured in P_n);
- α_n: normal clearance (measured in P_n);
- λ_s: edge inclination angle (measured in P_s).

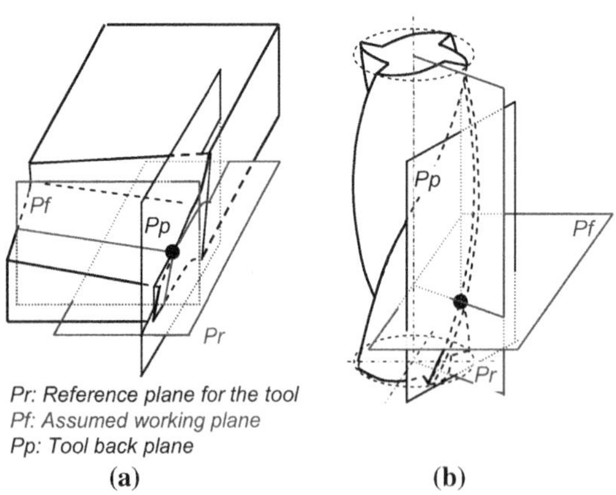

Figure 2.24 Tool-in-hand reference system for (a) turning and (b) milling tools

Figure 2.25 Tool-in-use reference system for a turning tool

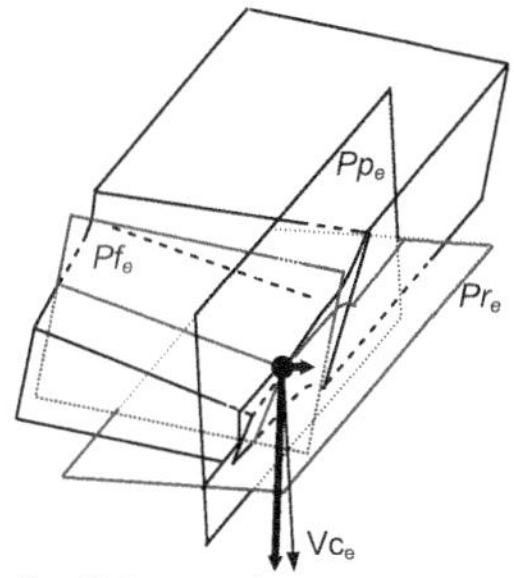

Pr$_e$: Reference plane
Pf$_e$: Working plane
Pp$_e$: Tool back plane

In Figures 2.27 and 2.28 the reference system is applied to and endmill and ball mill, respectively [11].

In the case of the tool-in-use reference system the main direction is given by the effective cutting speed (*i.e.*, the sum of the cutting speed and feed). Here three main planes are defined (see Figure 2.25):

- P_{re}: working reference plane, defined by the edge point and perpendicular to the effective cutting speed;
- P_{fe}: working plane, contains the cutting speed and feed vectors and is perpendicular to the reference plane;
- P_{pe}: tool back plane, perpendicular to P_{re} and P_{fe}.

2.5.1 Endmilling Tools

A main difference between endmills and ball-endmills is the helix angle variation along the cutting edge. Even though the possibility that the flutes of a ball-endmill may be with a constant helix angle, most of the tools in the market present constant lead, resulting in a variable helix angle. This is due to the usual grinding process applied for the fabrication of this kind of tool. Another consequence of the grinding process is that the normal rake and relief angles are made constant along the cutting edge.

Thus, the spatial generation of the edge is the result of projecting a cylindrical helix on a sphere perpendicularly to the axis of the tool direction. The resulting cutting-edge geometry is shown in detail in Figure 2.27. Here, cutting-edge angles have been measured following ISO nomenclature. The inclination angle, λ_s, is measured in the P_s plane (defined by cutting-edge discrete element AB and the cutting-speed vector V_c), and it is the angle formed by the cutting speed V_c and the cutting edge AB. Local helix angle, represented as i, is measured on the P_p plane (defined by V_c vector and Z-axis).

In cylindrical endmills (see Figure 2.26), the inclination angle of the edge coincides with the helix angle, since the tangent plane to the edge is always parallel to the tool axis.

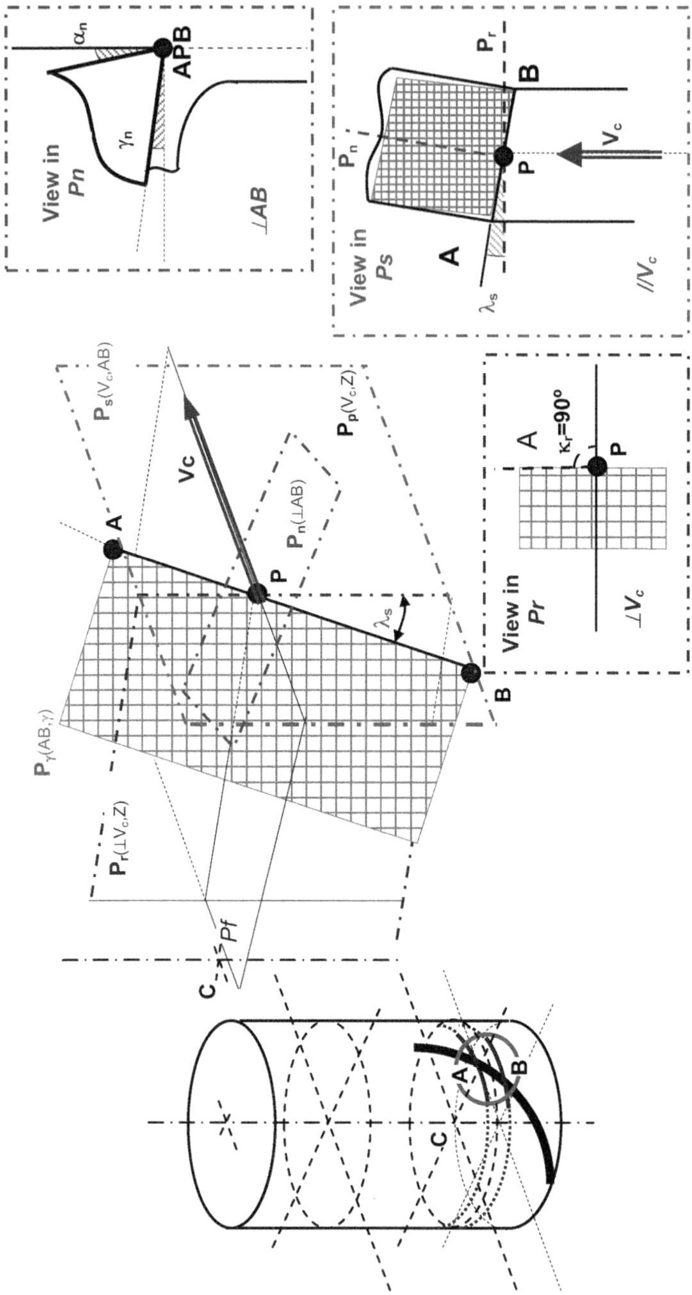

Figure 2.26 Geometry of an endmill

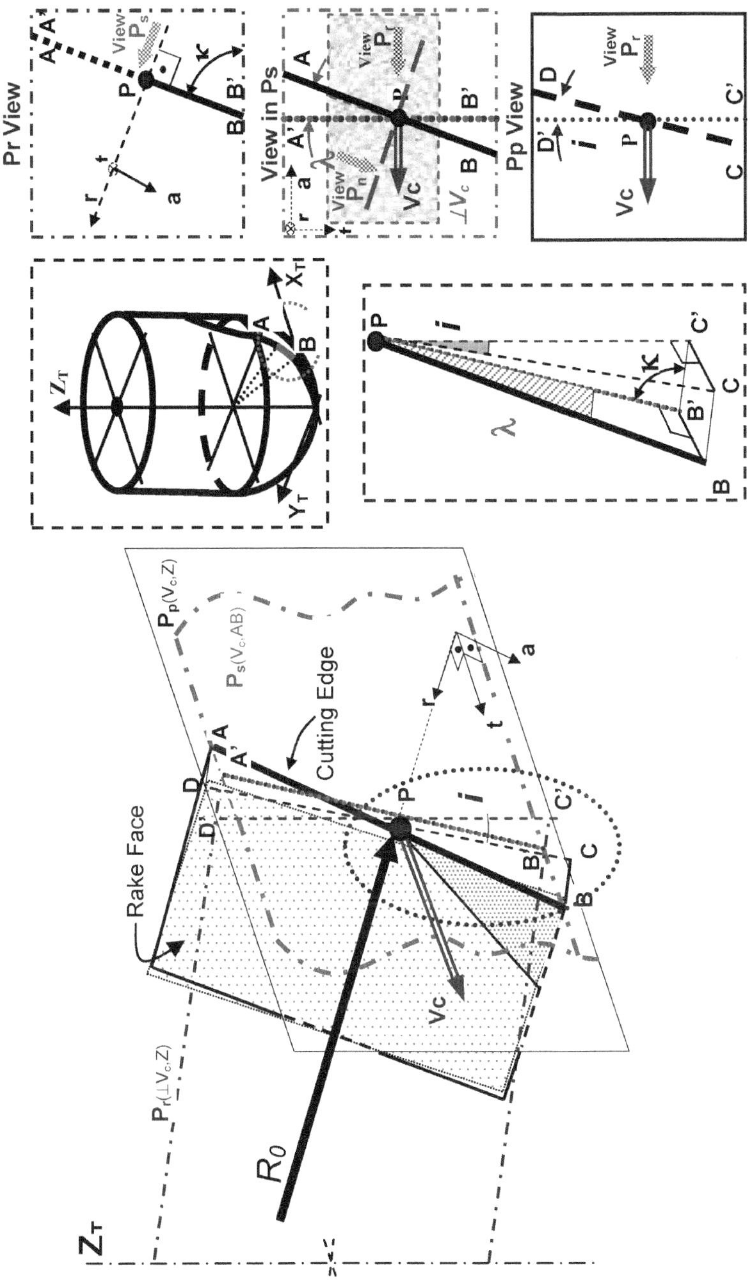

Figure 2.27 Geometry of a ball-endmill

2.5.2 The Rake and Clearance Angles

In oblique cutting, machining forces are composed of three components, instead of the two of orthogonal cutting. The geometry and angles of a cutting tool defines the values of each of the components, in addition to the characteristic shear angle Φ of the material and the inclination angle of the cutting edge with respect to the cutting speed, i. In Figure 2.28 the cutting-force components are described.

However, the most important concepts to evaluate in all the machining operations under a qualitative point of view are those presented in Figure 2.29: the clearance angle, the rake angle and the edge angle.

The rake angle shown in this figure is positive, the most common case, but in hard machining it could be negative, with the tool oriented towards the cutting-speed direction. Since clearance angle always must be positive to avoid rubbing on the part surface, a negative rake angle implies a very strong cutting edge, and therefore is recommended for very difficult-to-cut materials where cutting forces are much too high. Another aspect to bear in mind is the so-called edge radius, different from the corner radius (where the main and minor edges intersect). The edge radius is only few hundredths of a millimetre, and in some cases, as in PCBN inserts, a chamfer is produced instead of the rounded edge.

In milling tools both the mean and minor edges have a big influence on tool performance. Therefore the radial and axial rake and clearance angles must be considered. Several combinations could be used; in Figure 2.30 two of them are shown. The first situation in the figure is very aggressive, due to the both negative angles. The second situation is a good combination because the positive axial angle allows chip was evacuated out from the part surface, and the negative radial angle permits a very robust tool edge design.

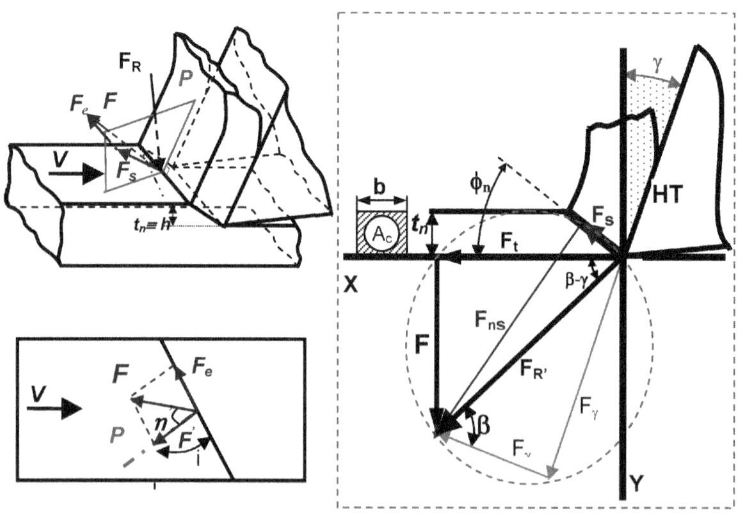

Figure 2.28 Geometry of the oblique cutting, with the decomposition of the cutting force in the main directions

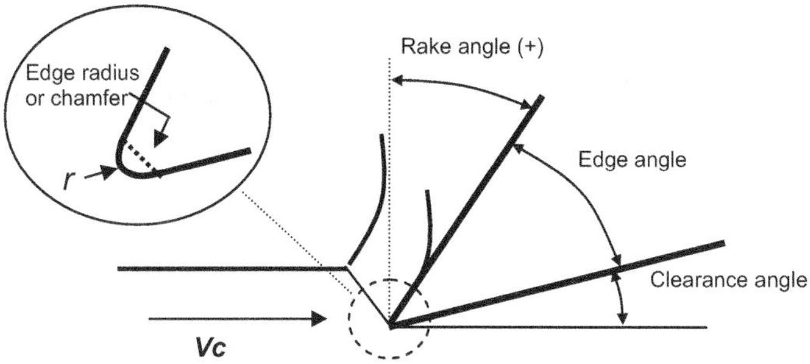

Figure 2.29 Basic angles for a rapid tool evaluation

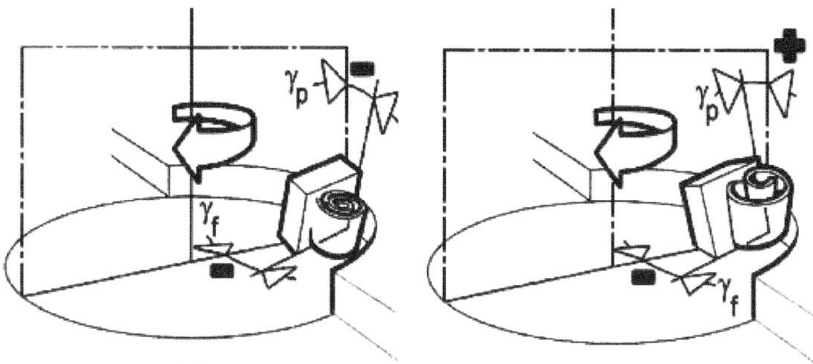

Figure 2.30 (a) Axial and (b) radial rake angles (courtesy of Sandvik Coromant)

2.5.3 *Position Angle*

The tool cutting-edge angle (κ_r) has a direct influence on chip thickness and therefore on the cutting-force components. At the same feed rate, decreasing the side cutting-edge angle increases the chip contact length and decreases chip thickness. As a result, the cutting force is dispersed on a longer cutting edge and tool life is prolonged.

Increasing the side cutting-edge angle increases chip width. Therefore, decreasing the position angle is recommended for:

- hard workpieces which produce high cutting temperature due to their high specific cutting forces;
- when roughing a large-diameter workpiece.

2.5.4 *Milling Tools for Several Applications*

Bearing in mind the geometry possibilities for milling tools and requirements derived from the workpiece hardness, shape and dimensions, big manufacturers of

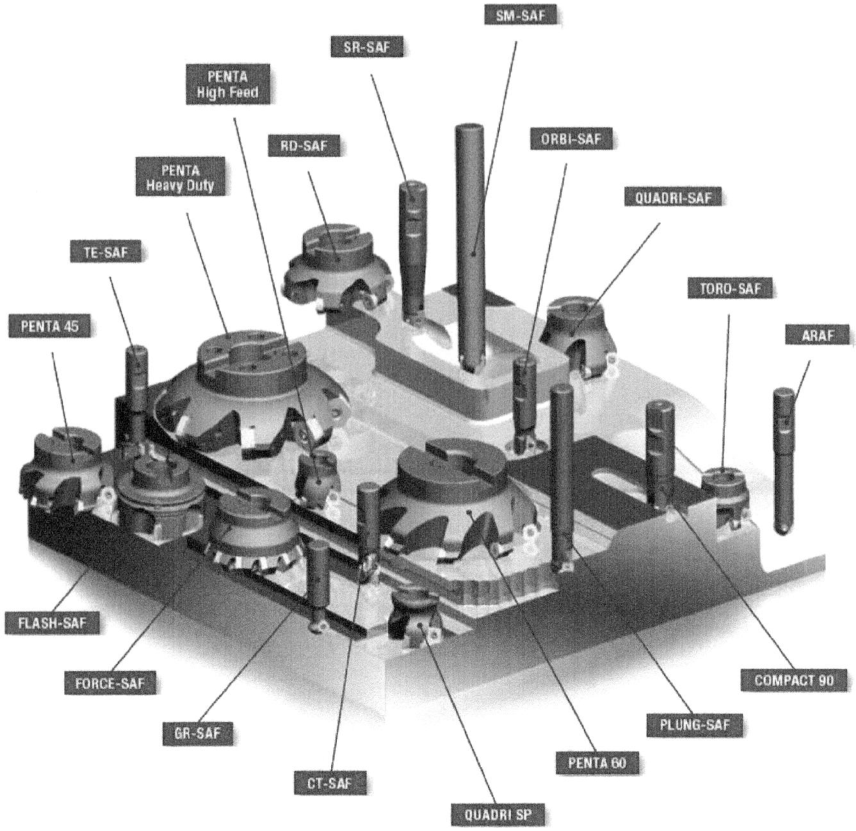

Figure 2.31 Milling tools supplied by Safety®

tools offer a complete catalogue of milling tools. In Figure 2.31 the milling options provided by the tool manufacturer Safety® are represented, including milling discs with inserts with different lead angles for facing, insert tools with lead angle 90° for slotting and shouldering, and ball-endmilling tools.

2.6 Hard Machining for Mould and Dies

Before the generalized use of high-speed milling, the usual technology employed in mould manufacture was a combination of conventional milling and electrodischarge machining [12, 13]. From 1997 to 1999 roughing and semi-finishing were usually carried out with conventional machines, with mould steel in a soft state before tempering. Subsequently, heat treatment was applied. After that, finishing was performed in high-speed machining centres. There were two reasons for such a sequence:

- Roughing, which is subject to few precision requirements, was done in machines which cost per hour one-fifth of high-speed machines. Moreover, tool wear was light because of the low hardness of the workpiece material.
- The most usual high-speed spindles available in those days were unable to deliver sufficient torque below 1500 rpm, making roughing impossible.

In 2000, technical variations made to high-speed spindles control resulted in an improved capacity to deliver enough torque even at low rotational speeds. In this manner, roughing in high-speed machines became possible, with a similar application practice to the conventional case. Therefore, a new procedure was defined starting directly from a block initially heat-treated, carrying out consecutively all operations in the same machine. The mean advantages of this simpler process was that less time was needed to launch a new mould, since between successive operations there was less time needed for set-up. At the same time, accuracy and reliability of workpiece also increased, due to the avoidance of workpiece zero setups between operations.

At present, the decision whether to use high-speed machines starting from tempered raw material, or conventional roughing of non-tempered steel followed by tempering and high-speed milling, depends on production costs and required lead times. But in all the cases finishing is performed by ball-end high-speed milling.

2.6.1 Ball-endmilling for Sculptured Surfaces

High-speed milling with ball-endmills is the basic technology for finishing complex surfaces, the final and high-added-value stage when complex forms are produced [12, 14–16].

In Figure 2.32 a typical ball-endmilling tool for finishing hard steels is shown. A four-flute geometry with a full cutting edge to the centre of the ball, in combination with an improved version of TiAlN coating (more than 3700 HV hardness) provides the necessary efficiency of cutting together with high heat and wear resistance.

This operation commonly involves the milling of a 0.3 mm allowance (as shown in Table 2.6), which is usually done using ball-endmills with diameter below 20 mm, due to the intricate shape details. Taking into account that slopes

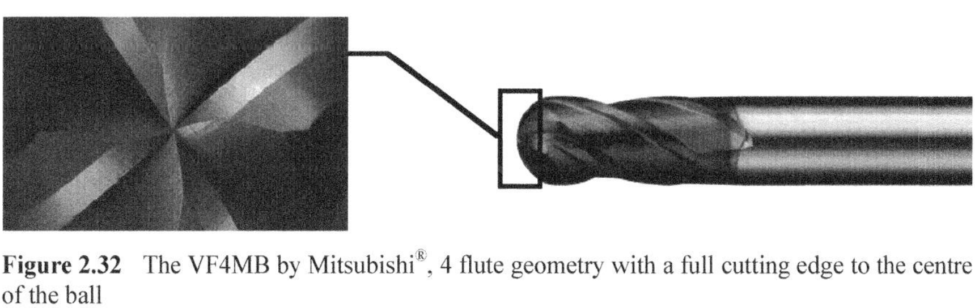

Figure 2.32 The VF4MB by Mitsubishi®, 4 flute geometry with a full cutting edge to the centre of the ball

Table 2.6 Cutting conditions recommended by Mitsubishi® for the VF4MB

Work material	Hardened steel (55HRC) W.Nr. 1.2344(H13)					Hardened steel (55 62HRC) X210Cr12				
R (mm)	%<15°		%>15°		Depth of cut (mm)	%<15°		%>15°		Depth of cut (mm)
	Revolution (min^{-1})	Feed rate (mm/min)	Revolution (min^{-1})	Feed rate (mm/min)		Revolution (min^{-1})	Feed rate (mm/min)	Revolution (min^{-1})	Feed rate (mm/min)	
R0.5	40,000	10,400	40,000	4,200	0.050	40,000	7,300	40,000	3,100	0.040
R1	40,000	12,500	39,000	6,100	0.090	40,000	10,400	24,000	3,100	0.080
R1.5	40,000	15,600	27,000	5,600	0.100	32,000	10,000	16,000	2,900	0.100
R2	32,000	14,100	20,000	4,700	0.120	24,000	8,100	12,000	2,500	0.100
R2.5	25,000	11,700	16,000	3,700	0.160	19,000	6,900	9,600	2,200	0.120
R3	21,000	10,900	13,000	3,400	0.200	16,000	6,200	8,000	2,100	0.160
R4	16,000	8,300	10,000	2,600	0.240	12,000	4,700	6,000	1,600	0.160
R5	13,000	6,800	8,000	2,300	0.400	10,000	4,200	4,800	1,200	0.160
R6	9,000	4,700	6,000	1,700	0.400	7,000	2,900	3,600	940	0.240

commonly found in sculptured forms go from 0° to 90° and that effective cutting speed must be between 300 and 400 m/min – this is the maximum recommended for the current carbide tools coated with AlTiN – the spindle rotational speed must be over 15,000 rpm. This means that high-speed spindles must be used. Nowadays, the maximum rotational speed of industrial electrospindles is around 20,000–25,000 rpm, with power ranging from 14 to 20 KW.

On the other hand, for this rotational speed, and bearing in mind a recommended feed of around 0.07–0.1 mm/tooth, the maximum linear feed is 10–15 m/min. These values can be obtained by typical linear ball screws connected to synchro-

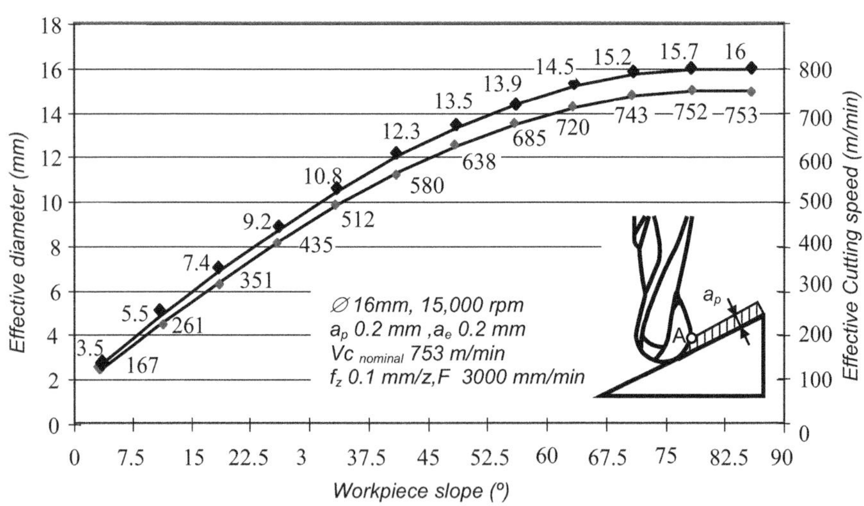

Figure 2.33 Effective tool diameter and cutting speed for different surface inclinations, using a ball-endmilling tool. *A* is the point of maximum cutting speed

nous motors, or by linear motors. The machine axis control at this feed is not a problem for current numerical controls.

In Figure 2.33 values of maximum cutting speed at the effective diameter are shown for a ∅16 mm integral ball-endmilling tool. Integral carbide tools are more used than insert tools for finishing.

2.6.2 *Five-axis Ball-endmilling*

In five-axis ball-endmilling, two additional orientation axes added to the machine allow the machining of very complex parts, which cannot be machined using three-axis machines [17].

Otherwise, cutting speed is zero at tool tip, making the tool cutting very unfavourable. This is because when ceramics or PCBN tools are used, typical failure is the fragile breakage of the tool tip. With five axes, milling can be performed avoiding the tool tip cutting.

Moreover, tool overhang, necessarily large when deep cavities are machined, can be reduced using five-axis milling. Therefore, tool stiffness is higher, which increases machining precision and reduces the risk of tool breakage. Tool stiffness is directly related to the tool slenderness factor L^3/D^4 [15, 18], so a tool length (L) reduction dramatically reduces tool deflection and the lack of precision due to this effect.

2.7 Toolholders and Tool Clamping Systems

The assembly of tools in machining centres is a key factor for obtaining parts with high dimensional accuracy and surface quality. Moreover, the performance of a tool can be significantly influenced by the quality of the clamping system to the machine. In general, the use of clamping systems as rigid as possible is recommended in order to reduce the toolholder–tool deflection and provide a secure holding system for the tool for high-performance machining conditions. In addition, a rigid clamping system is the basis for an accurate and precise tool–toolholder–spindle assembly. However, on the other hand, it is necessary to provide a simple tool-change solution to obtain minimum chip-to-chip times. Currently more different tools, specifically designed for single operations, are used to carry out individual operations. Therefore, the number of tool changes required to machine a complete part is higher than traditional machining strategies. The answer of machine-tool builders to this requirement has been the development of new methods for a quicker and more precise tool change, with systems capable of managing hundreds of different tools and with chip-to-chip times even below 0.7 s.

In general, toolholders must achieve the following capabilities:

- assembly and disassembly must be simple;
- allow automatic tool change (ATC) commanded from the CNC;
- maximum coaxial accuracy on the tool–toolholder–spindle assembly;
- maximum stiffness of the complete system;
- maximum torque transmission from the spindle to the tool.

It is important to take into account that a finely designed tool clamping system would not improve the behaviour of the tool; however, an incorrect clamping system would reduce tool life significantly.

The most common solution is the introduction of an intermediate component which on one extreme holds the cutting tool and on the other is fixed to the machine-tool spindle or turret. Therefore, there are two mechanical interfaces between the tool tip and the machine-tool spindle: *tool–toolholder* clamping and *toolholder–spindle* (or turret for lathes) clamping.

2.7.1 Toolholders for Turning Operations

Tool holding systems for turning operations are relatively simple, since in lathes the tool is fixed to the machine turret rigidly. The most common system is the use of the standard DIN 69880 (VDI) clamping, which consists of bars of cylindrical section attached to the turret lathe, with a serrated shape. However, there are solutions that allow more flexibility and an easier assembly of different tools. These systems are based on a specially developed joint between tool and toolholder. The most extended system is the Capto®, originally developed by Sandvik Coromant™ but recently out of patent. Figure 2.34 shows both solutions: a toolholder based on the DIN 69880 standard and the Capto system. The former also presents the VDI serrated-shape interface of toolholders for indexable turrets.

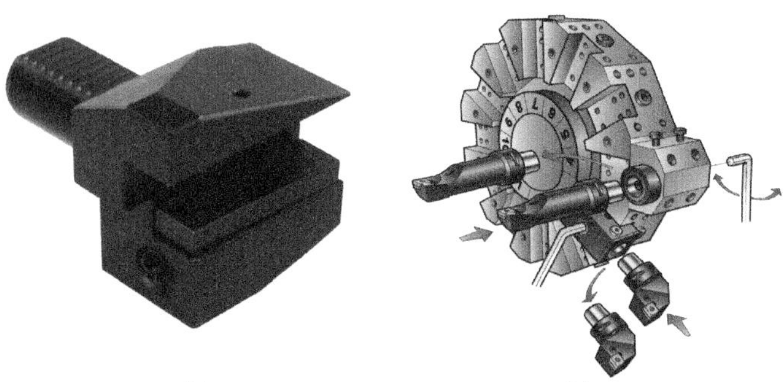

(a) (b)

Figure 2.34 (a) DIN 69880 (VDI DIN69880) toolholder, and (b) Capto® system for turning tools

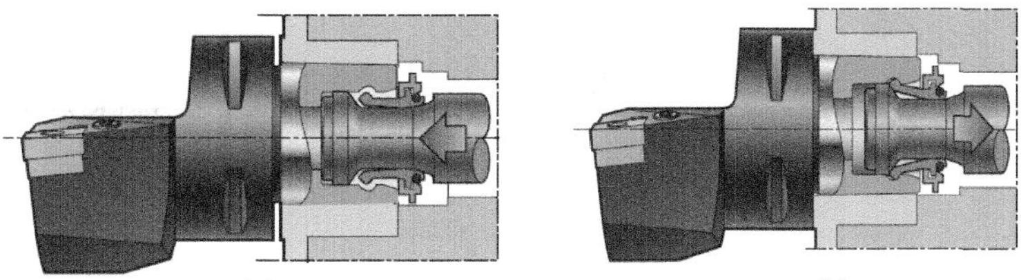

Figure 2.35 Capto system scheme: (a) unclamped position, and (b) clamped position (courtesy of Sandvik Coromant)

The Capto clamping system is based on the double interplay between tool and holder surfaces, both external and internal surfaces. While the external surfaces are based on a combination of polygonal and radius-shaped design and provide torque transmission, the internal clamping device allows easy clamping and unclamping of the tool. The system is based on a segmented expandable bushing (see Figure 2.35) in the clamping unit, and lips on the outer periphery of the segments lock into an inner groove on the cutting unit, clamping the two components together. In the unclamped position, the drawbar is in the forward position; the forward ends of the segmented bushing move towards the centre line of the coupling. The diameter is reduced and the lips on the outer edge of the bushings disconnect from the inner groove of the cutting unit. The drawbar pushes the cutting unit out. In the clamped position, the drawbar is in the retracted position; the forward ends of the segmented bushing are forced outwards away from the centre line of the coupling by the shoulder on the drawbar. The lips on the outer edge of the bushings lock into the inner groove of the cutting unit which is pulled into its working position.

In addition to turning, Capto is currently a solution for milling tools as well, being in use in multitasking operations (see last section of this chapter).

2.7.2 Toolholders for Milling Operations

The role of the toolholders in milling operations is similar to other machining operations, since high stiffness, holding reliability and tool position accuracy are required. However, today milling processes apply high spindle speeds (up to 40,000 rpm), which produce high centrifugal forces where the rotational system presents unbalanced elements. This fact has forced designers to rethink aspects such as the joint between toolholder and machine-tool spindle or the requirement for toolholder balance. Therefore, the conventional toolholders for milling operations, based on the single lateral contact face of the tapered shank, are being substituted by systems with a double contact face: lateral and perpendicular to tool axis (Figure 2.36).

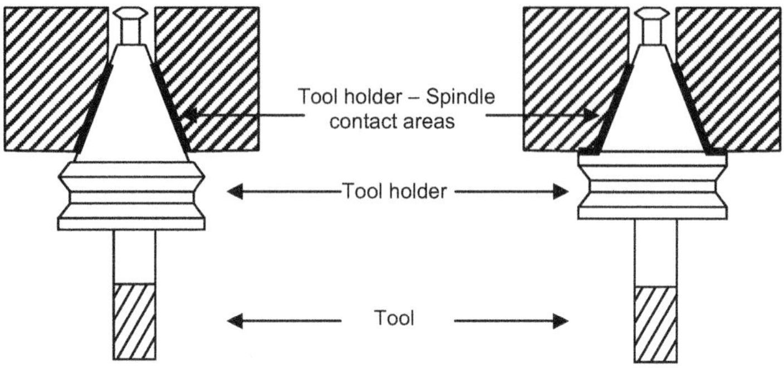

Figure 2.36 Toolholder for milling: single-contact system (*left*), and double-contact system (*right*)

Single-contact toolholders have been used since the development of ATC. The most common is the ISO-7388, which consists of a tapered toolholder to be inserted into the machine spindle. These toolholders, known simply as ISO tooling, can be used reliably up to 6,000–8,000 rpm. However, in the last 15 years the development of new tool materials, spindle technology (including electrospindles) and high-performance machine tools has allowed the emergence of HSM. In this technique, cutting speed is increased by more than five times the conventional speed. Therefore, the rotation speed of the milling tools has to be increased too, up to 40,000 rpm in some cases.

If rotation speed is higher than 8,000 rpm, centrifugal forces are relevant and the single-contact system loses joint stiffness rapidly. One of the main problems of the ISO toolholders comes from the clamping system. ISO toolholders are clamped by a mechanical system that pulls up the toolholder and it is released by an actuator (both hydraulic and pneumatic). If spindle speed increases, centrifugal force can cause lateral expansion of the spindle axis, while the clamping system continues pulling up the holder. Thus, the spindle is pulled inside the spindle nose, causing inaccuracies and it is even possible for it to be stuck in the spindle nose.

Moreover, the mass offsets of the rotating tool–toolholder system with respect to the rotation axis cause unbalancing forces, which depend on the square of the rotating speed. Therefore, the same tool–toolholder system (with the same unbalanced elements) rotating from 4,000 rpm to 20,000 rpm increases the unbalancing forces 25 times. As a consequence, new toolholders and balancing systems were introduced to reduce the centrifugal force effect and unbalancing problems.

In order to reduce these problems, other clamping systems have been developed, such as the HSK.

2.7.2.1 HSK Toolholders

HSK is the acronym of a new standard tooling interface for milling toolholders; it basically means "hollow shank tooling". It was developed in Germany in the late 1980s and rapidly became a standard in Europe. Actually it is widespread in Asia and the USA as well. The standard references for HSK tooling are DIN69893

(a) (b)

Figure 2.37 (a) Heat shrink holder HSK63A, and (b) detail of HSK63 A holder

and spindle receivers DIN69063. These standards were introduced as non-proprietary solutions and describe the specifications for HSK.

The HSK system presents some advantages with respect to the ISO system. One of the most relevant is that HSK present a double-face contact system. This difference is a key factor in high-speed machining operations since the reference surface for the toolholder is the spindle nose [19]. In addition, the dual contact systems achieve better repeatability on automatic tool changes.

Another important difference is the clamping system. HSK toolholders are fixed by a segmented expandable bushing driven by a drawbar. The segments are inserted in a cup-shaped (Figure 2.37) hollow machined in the toolholder. Therefore, if spindle speed increases, the centrifugal force expands the segments and consequently the clamping force is increased too. This capability allows for more aggressive cutting conditions; in addition it provides greater rigidity and accuracy than systems based on ISO holders.

Machines using ISO holders are also more sensitive to chatter than those using HSK because the junction between toolholder and spindle is not as rigid. The lower rigidity of this union drops the natural frequency of vibration and limits the material removal rate.

There are different types of HSK holders. They are defined by two or three digits and a letter, for example HSK-63 A (one of the most common in use). The figure gives the outer diameter of the plate that sits on the spindle nose. The letter indicates the type of holder depending on various factors such as length, guidance systems, *etc*. In general, the most usual types are:

- A: General type, in use in more than 95 % of machines.
- B: It has a larger flange than the A type. It is used for more aggressive cutting conditions.
- E and F: Same as A and B but without marks and guidance systems for enhanced balance.

As noted, HSK has great benefits, but there are some disadvantages with respect to the ISO clamping system. First, the HSK tooling is more complex and expensive. Second, HSK is very sensitive to the presence of particles such as chips or grease. Moreover, there can be chips in the hollow where segments guided by the drawbar have to fix the holder to the spindle. This sensitivity to impurities requires extreme care during tool changes, and the usual solution is to inject pressurized air into the spindle nose and the holder before each tool change.

2.7.2.2 Other Toolholder Systems

There are other types of holders widely used in milling and drilling operations. These systems are based on dual-flange holders and are either V-flange or BT-flange, depending on the precise flange configuration. V-flange toolholders are often referred to as CAT tooling (from Caterpillar), because the initial design was developed about 30 years ago by engineers at Caterpillar Tractor Co. working in conjunction with machine-tool builders. The design eventually became a national standard, and the majority of toolholders currently in use in the US are CAT style. Japanese and European applications, on the other hand, may use BT-flange holders, described in the Japanese standard JIS6399 (MAS-403). Both systems use single-contact surface systems, so similar problems with ISO systems have to be expected if spindle speeds increase over 8,000 rpm. BT holders actually present a version with double-contact for high-speed milling.

Another holder type is the BIG-PLUS® system, with simultaneous dual contact between the machine spindle nose and toolholder flange face. This system is based on the most currently available standards for JIS-BT, DIN69871 and the CAT-V flange tooling and actually is licensed by more than 100 machine-tool and spindle manufacturers.

2.7.3 Tool–Toolholder Clamping Systems

As mentioned above, there are two different joints between the machine spindle and the tool tip: first, the toolholder and machine tool spindle joint, which has been described in the previous section; second, the joint formed by the toolholder and tool. The connection between tool and toolholder has to satisfy the same requirements of accuracy, stiffness, torque transmission and interchangeability as the spindle-shank one. Therefore, different mechanical solutions have been developed to perform these specifications. Obviously, each solution presents advantages and disadvantages with respect to others and all of them are being used nowadays.

Basically there are three types of rotary tool clamping systems: collet chucks, hydraulic holders and shrink-fit holders.

2.7.3.1 Collet Chuck Tool Clamping Systems

It is the most common solution, based on introducing the tool into a segmented collet which is inserted into the holder (Figure 2.38). The clamp force is achieved by a nut that presses on the segments of the collet. The collet segments are designed to increase the flexibility of the collet and to obtain a uniform pressure on the contact surfaces between tool and collet, and collet and holder.

The collet system is valid for most of the high-speed machining operations and it is the most economical solution. Another advantage of this system is that it may have different collets for a single holder, so different diameter tools can be used in the same holder.

In terms of precision, high-quality collets can obtain a run-out near 7–8 μm at 25 mm from the spindle nose. These results can be achieved with high-quality mechanical holders and collets, manually adjusted.

However, some applications require lower run-out values. Moreover, the stiffness of the clamping system cannot be enough. In these cases, the holders should use hydraulic or shrink-fit tool clamping. Both systems provide more rigidity and precision.

2.7.3.2 Hydraulic-expansion Tool Clamping Systems

Hydraulic-expansion holders clamp the tool through a hydraulic system (Figure 2.39). There is a metallic membrane surrounding the tool shank. The membrane

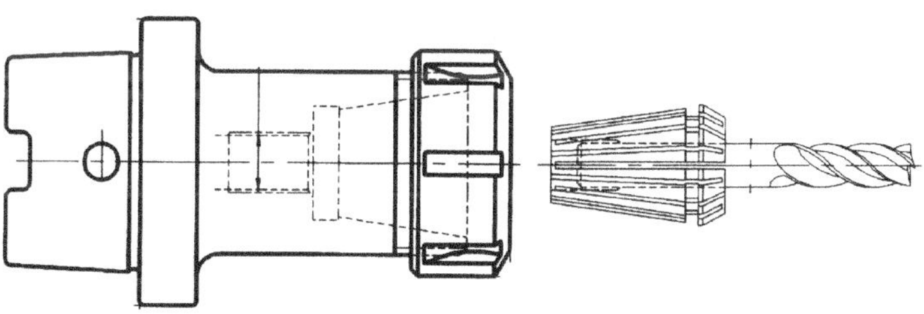

Figure 2.38 Collet-based toolholder (courtesy of LAIP®)

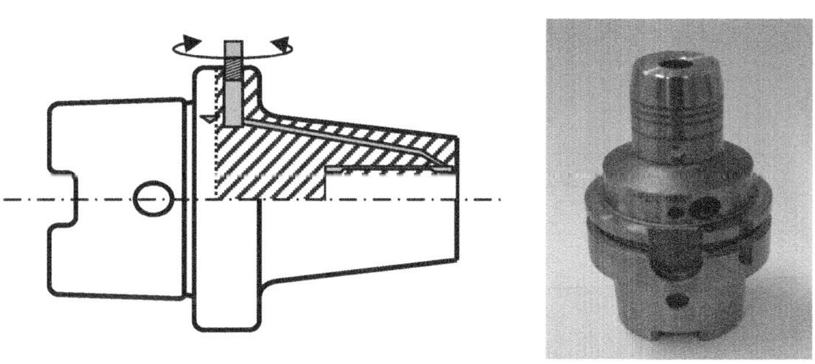

(a) (b)

Figure 2.39 (a) Hydraulic-holder scheme, and (b) hydraulic HSK63 A holder

is surrounded by a fluid deposit; the fluid pressure can be incremented by a screw, which moves as a piston. Therefore, the tool is clamped by the membrane, which transmits the pressure of the fluid to the toolholder uniformly. Since all the fluid is inside the holder, chips or cutting fluid do not affect the toolholder.

The main advantage of these systems is the high accuracy of the tool and toolholder union. Some commercial suppliers guarantee run-out values below 2.7 μm measured 30 mm below the spindle nose on 12 mm diameter endmills. These run-out values make these holders suitable for ultra-high-speed machining operations (more than 30,000 rpm) and for high-accuracy operations with small endmills. Moreover, material removal rate can be increased since the tool is perfectly balanced and there is no misalignment between spindle and tool axes.

On the other hand, there are two major drawbacks. First, the cost of this type of tooling can be up to five times higher than conventional mechanical collet tooling. Second, each holder must be used only for a tool diameter, since the membrane is adapted for a specific shank diameter. Therefore, a different holder is needed for each tool shank. There are some solutions for this problem, usually based on using additional membranes that can be inserted in the holder. However, this solution increases the run-out values up to 1 μm for each additional membrane.

2.7.3.3 Heat-shrink Tool Clamping Systems

Heat-shrink toolholders provide high accuracy and minimum run-out at a reasonable cost. Unlike hydraulic ones, there are no internal systems to bring pressure to hold the tool. Instead, the holder consists of a monolithic element with a precision hole where the tool is inserted.

At room temperature, the hole is slightly smaller than tool diameter. Using an external heater, the cone is heated and the tool housing hole expands. The heater

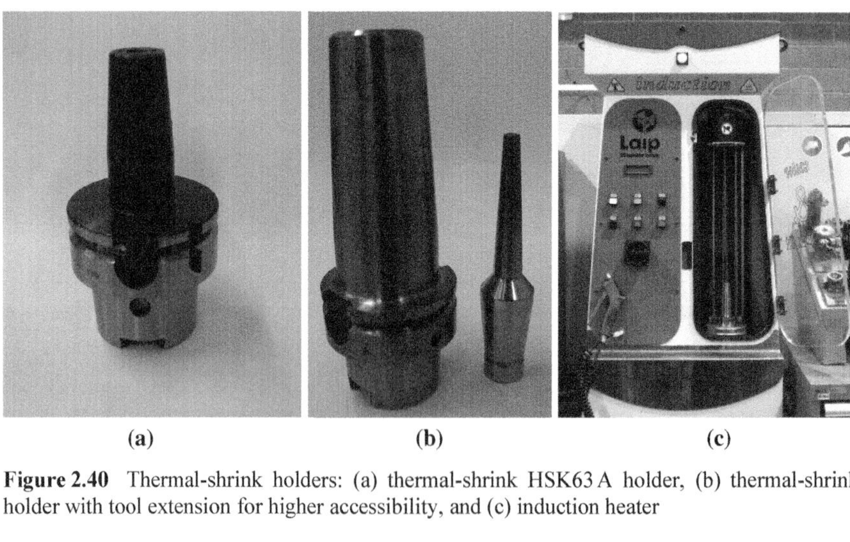

(a) (b) (c)

Figure 2.40 Thermal-shrink holders: (a) thermal-shrink HSK63 A holder, (b) thermal-shrink holder with tool extension for higher accessibility, and (c) induction heater

can be as cheap as a hot-air heater, but more sophisticated heaters based on induction are being used industrially (see Figure 2.40). Once the hole has expanded, the tool is introduced and the holder is cooled again to room temperature. When the holder recovers its original dimensions, tool is clamped strongly. This method provides excellent stiffness and minimum run-out (values are comparable to hydraulic holders). Furthermore, there are no additional items such as screws, nuts, *etc.* to hold the tool, so an excellent balancing is achieved.

On the other hand, it is necessary to have a holder for each different tool diameter, which may introduce an added cost.

2.7.3.4 Toolholder Balancing

As mentioned, the toolholder balancing is a key factor in high-speed machining. The unbalancing of a system depends on the unbalanced mass and the position of its centre of gravity about the rotation axis. Since it is impossible to get a perfectly balanced system, the objective is to reduce the unbalance to the minimum value.

The tool–toolholder–machine-tool spindle unbalancing can be originated by a combination of factors:

- asymmetric elements in the toolholder, such as screws, wedges, grooves, *etc.*;
- tools with asymmetric shapes (Weldon holder, drills with one cutting edge, *etc.*);
- imperfections in the holder, tools or collets.

Thus, balancing is defined as the amount of mass multiplied by the eccentricity of the mass. The problem is not the unbalance itself, but the combination of the unbalance with high spindle speeds. The force due to the unbalance can be calculated as:

$$F_{\mathrm{U}} = U\left(\frac{S}{9550}\right)^2$$

where F_{U} is the unbalance resultant force measured in newtons, U is the system unbalancing measured in gram-millimetres and S represents the spindle speed measured in revolutions per minute.

If the unbalance is about 6 to 8 g mm, the force due to unbalance at 18,000 rpm could be higher than the cutting force, especially in finishing operations. Therefore, in order to limit the unbalance effects, tooling manufacturers use ISO 1940-1 to establish the degree of admissible unbalance of the tooling. This standard establishes different *G classes*. The lower the G class, the better balanced is the tooling. Many manufacturers are producing tooling class G1.0 to G2.5. This G class gives the maximum allowed unbalance using the formula

$$U = \frac{9553mG}{S} \tag{2.3}$$

where U is the admissible balance measured in gram-millimetres, m is the total mass of the system measured in kilograms and G is the G class of the system following the ISO 1940-1 standard.

2.8 New Techniques for Hard Machining

The development of new tools, on one hand, and the new multitask machines, on the other, has allowed the proposal of new machining techniques for roughing, semi-finishing and finishing. The main purpose of them is to avoid machining vibrations and to take advantage of the stiffer direction of machining centres, that is, the spindle axis.

2.8.1 High-feed Milling

High-feed milling is a roughing technique that works with cutters and inserts designed specifically for the technique. Inserts typically feature large sweeping radii and positive rakes (see Figure 2.41). The high-feed method takes advantage of small setting angles (55° or less). This produces a minimal radial and a maximum axial cutting force. As a matter of fact, the cutting forces are directed towards the machine spindle in the axial direction. This is the stiffer direction of the machine, which reduces the risk for vibrations and stabilizes machining.

This allows for higher cutting parameters even when machining with a large overhang. Therefore, instead of cutting with greater depth, it does the opposite: it pairs shallow depth of cut with high feed per tooth, in some cases higher than 1.5 mm.

At the same time the axial depth of cut is very small, leading to a near-final shape in the case of complex surfaces. Consequently semi-finishing operation is eliminated. This greatly reduces the machining time in the case of moulds or dies.

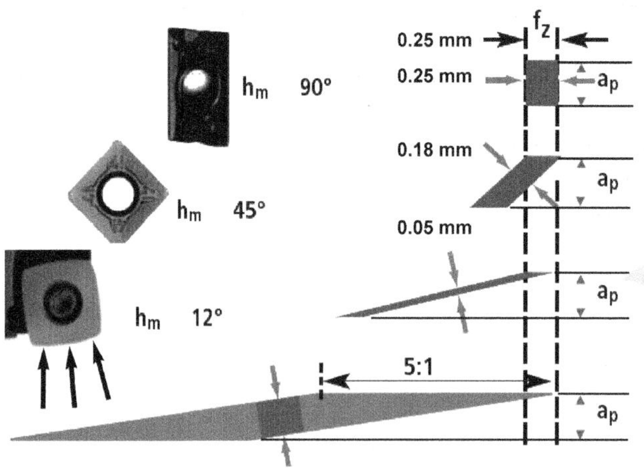

Figure 2.41 Effect of the tool position on the chip section, the basis of the high-feed milling technique (courtesy of Stellram®)

High-feed milling inserts can make facing, ramping, helical interpolation and plunging operations. The interpolation capability of modern CNC machines makes it possible for a small tool to mill out a much larger hole or pocket by ramping. The tool ramps from one level of passes to the next within the feature, or it follows a helical path at a continuous angle all the way down to the feature's depth. Ramping angle in penetration depends on the clearance between insert and part surface and therefore indirectly depends on the insert size, being higher for the smaller inserts.

Inserts can present three, four, five or six cutting edges. Thus, Safety uses pentagonal inserts having five cutting edges (see Figure 2.42 and Table 2.7). Inserts for use in the V556 tools encompass two geometries for roughing and finishing, respectively, and four grades, including VP5020 and VP5040 multilayer PVD TiAlN/TiN coated grades for general applications, a TiN/Al2O3/TiCN CVD VP5135 coated

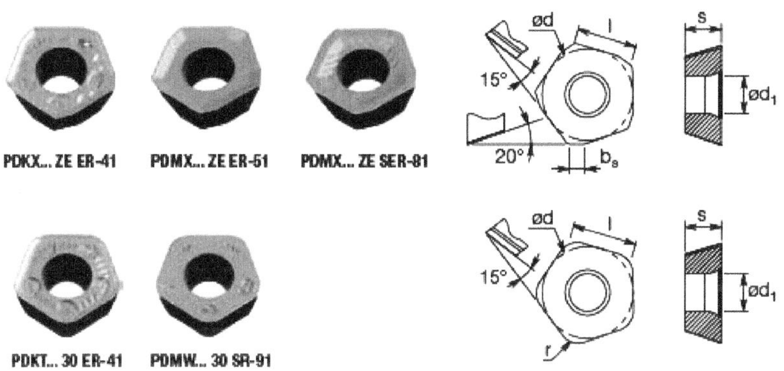

Figure 2.42 The Penta-edge insert for high-feed milling of Safety®

Table 2.7 Recommended values for the Penta high-feed inserts by Safety® (for a tool life of 15 min)

	Material	Hardness	Cutting speed v_c (m/min)		Tooth feed f_z (mm)			
					Face milling with $a_e = D$ & $a_p = 1$ mm		Ramping with $a_e = D$	
			Grades		Geometries		Geometries	
			5020	5040	51	81	51	81
P	Non-alloy steel	< 150HB	180 (130 - 220)	140 (100 - 160)	0.5 - 1.5	1.0 - 2.5	0.5	0.8
	Low-alloy steel	150 - 280 HB	140 (110 - 170)	110 (90 - 130)	0.5 - 1.0	0.8 - 2.0	0.5	0.8
	High alloy steel	180 - 280 HB	110 (90 - 130)	100 (80 - 120)	0.5 - 1.0	0.8 - 2.0	0.5	0.8
M	Stainless steel	< 270 HB	120 (100 - 140)	110 (90 - 130)	0.5 - 1.0	0.8 - 1.5	0.5	0.8
K	Grey cast iron	< 350 N/mm²	250 (200 - 300)	-	0.5 - 1.5	0.8 - 2.5	0.5	0.8
	Nodular cast iron	< 800 N/mm²	200 (200 - 250)	-	0.5 - 1.5	0.8 - 2.0	0.5	0.8
N	Non-ferrous metal	90 HB	1000 (700 - 2000)	-	1.0 - 3.0	-	1.0	-
S	Heat resistant alloys	-	40 (30 - 50)	-	0.5 - 1.0	-	0.5	-
	Titanium	350 HB	35 (25 - 45)	-	0.5 - 1.0	-	0.5	-
H	Pre-treated steel	30 - 45 HRc	120 (80 - 160)	100 (80 - 120)	0.5 - 1.0	-	0.5	-
	Hardened steel	45 - 55 HRc	70 (50 - 90)	60 (50 - 80)	-	-	-	-

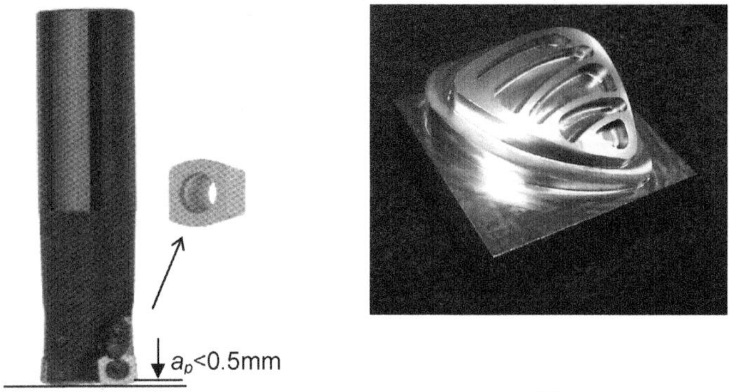

(a) (b)

Figure 2.43 (a) Hitachi Alpha Plus tool, and (b) small mould made by high-feed milling and high-speed milling (tool ∅ = 25 mm, f_z = 2.2 mm/z, N = 1700 rpm, a_p = 0.7 mm, a_e = 18 mm)

grade for tough machining, and VP1120 abrasive-resistant grade ideal for grey and ductile cast irons.

Four-sided inserts produce a side wall that is close to a square profile, this being the main advantage of these inserts.

In Figure 2.43 a small mould in a 35 HRC steel is presented as an example, being machined from a initial raw block of 130 × 90 × 90 mm. In the high-feed roughing the feed per tooth was 2.2 mm, a_p 0.7 mm and a_e 18 mm with a Hitachi Alpha Plus™ tool of ∅25 mm. The component was finished in only 17 min in a roughing-finishing sequence. High-speed finishing was performed at 20,000 rpm with a ball-endmilling tool.

2.8.2 Plunge Milling

This is a high-performance roughing technique in which a milling tool is moved multiple times in succession in the direction of its tool axis or of its tool vector into the material area that is to be removed, forming plunge-milling bores. The bores are superposed to eliminate the material of a pocket or zone.

This technique is also referred to as *milling in the Z-axis*; it is more efficient than conventional endmilling for pocketing and slotting difficult-to-machine materials and applications with long overhangs.

The machining parameters depend on the insert size, the tool overhang and the tool diameter. When a tool overhang of ∅6 mm is used, the usual step between two bores must be lower than 0.75 ∅. The radial depth of cut is 1 mm less than the radial length of the insert edge. If overhang increases the step must be reduced.

The advantages of the plunge-milling technique are:

- reduction by half in the time needed to remove large volumes of material;
- reduced part distortion;

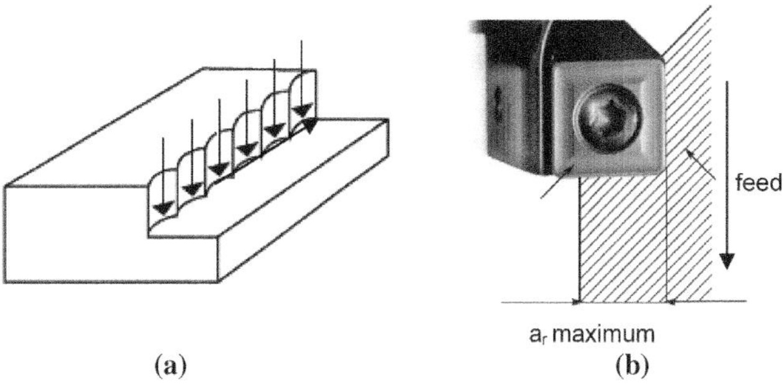

Figure 2.44 Plunge milling: (a) general procedure, and (b) maximum radial width of cut (courtesy of Stellram®)

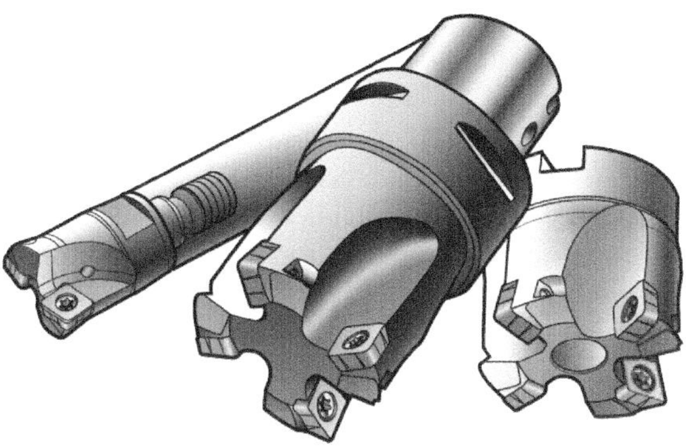

Figure 2.45 Sandvik® Coromill 210 is suitable for both high-feed and plunge milling

- lower radial stress on the milling machine, meaning spindles with worn bearings can be used to plunge mill;
- long reach, which is useful for milling deep pockets or deep side walls.

Plunge milling is recommended for jobs such as roughing cavities in moulds and dies. It is recommended for aerospace applications, especially in titanium and nickel alloys.

Inserts specifically for plunging are available for roughing and semi-finishing, but inserts suitable for high-feed milling can be also used for this technique. In Figure 2.44 the insert of system 7791VS by Stellram® is shown, specifically designed for this application. In Figure 2.45 some milling tools for both feed and plunge milling are shown.

2.8.3 Turn Milling and Spinning Tool

Two operations were recently developed for application in the new-generation multitask machines, face turn milling and the spinning tool.

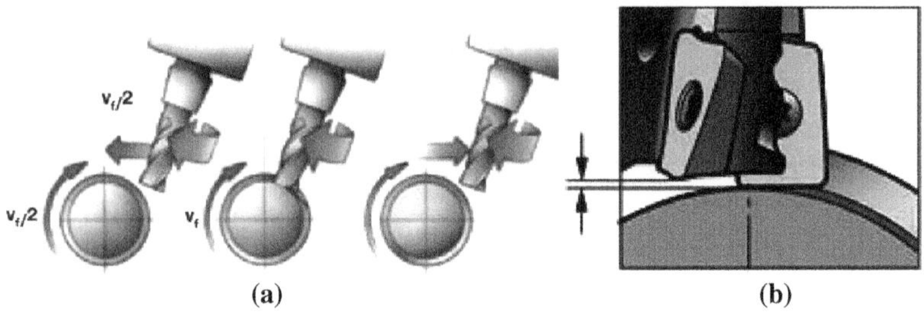

Figure 2.46 (a) Face turn milling (courtesy of Sandvik®), and (b) detail of the wiper insert

In face turn milling one wiper insert is used to generate the straight-line contact between the cutter and the machined surface in order to create the cylindrical part of the component. A wiper insert (see Figure 2.46) is one that follows along behind the cutting edge, extending just a little farther into the material to smooth out the freshly machined surface, avoiding the usual scallops of milling surface patterns.

The rotational speed of parts must be equal to the recommended feed per tooth. Basically it is a face-milling operation where feed is applied in a rotational way by the *C*-axis of the lathe. The basic parameters for milling can be directly applied to this practice.

As main advantage the chip control [20] offered by interrupted cutting can be highlighted in comparison with the long chips of turning. Other applications and advantages can be regarded:

- Turning tools tend not to do well in interrupted cutting, but a milling tool can fare much better. A milling cut is already an interrupted cut by definition. In the region of the workpiece where the cut becomes interrupted, it may make sense to switch from turning to turn milling.
- When the turned part is long, slender and not braced in the middle, turn milling may prevent it from deflecting.
- In a hard-to-machine metal, a single turning insert might not be able to deliver enough tool life to last to the end of the cut. A milling tool can cut longer, because it has multiple inserts to divide the load.

The radial (*X*-axis) motion of the milling cutter can be coordinated with the rotation of the workpiece to machine profiles other than perfect circles. Sandvik itself uses this technique to rough-machine the three-face, tapered shape of its Capto toolholders. The same principle, the milling cutter moving in and out while the workpiece turns, can also be used to generate off-centre features without having to change the setup. The off-centre pin on a crankshaft could be an example of this.

Y-axis motion is needed because the milling cutter has to do most of its cutting off-centre. The tool cannot machine the part to its final shape and dimensions when it is on-centre, that is, when the tool centre is located on the cylindrical part axis. In this case the endmill would cut with the centre point and not on its edges.

(a) (b)

Figure 2.47 Spinning tool, developed by Mori Seiki® (a) and Kennametal® (b)

Therefore the tool centreline should be offset from the work's axis of rotation by a quarter of the cutter diameter to cut properly. Using this approach, the problem appears when the tool reaches a shoulder: a rounded corner is produced by the off-centred endmill. To achieve a sharp corner, the cutter must take a second pass. The offset is eliminated, so the tool moves back to the on-centre position in *Y*. This second pass cleans the corner material away.

Spinning tools are another approach, where the cutting speed is the sum of the rotational speed of the cylindrical part and the milling movement at high rotational speed. This new cutting technology uses a specialized insert – similar in design to a round, or full-radius insert – mounted at the bottom of a cylindrical tool shank (Figure 2.47). Designed to distribute heat and wear more effectively than a single-point lathe tool, the spinning-tool technology can increase productivity by up to 500 % and tool life by up to 2,000 %.

This approach competes technically against traditional turning with single-point tools where the cutting force produces a torque and bending on the tool and gives rise to vibrations. But in the case of the spinning tool, most of the cutting forces are directed axially into the spindle and hence significantly reduce vibrations. The spinning tool can also cut in a back-and-forth motion, and this capability was also demonstrated on taper and arc shapes.

2.8.4 Trochoidal Milling

A trochoidal toolpath is defined as the combination of a uniform circular motion with a uniform linear motion, *i.e.*, toolpath is a kinematics curve so-called trochoid (Figure 2.48). Light engagement conditions and high-speed milling are

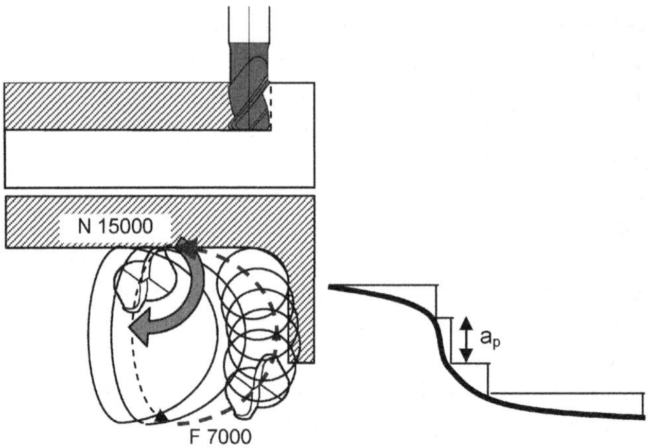

Figure 2.48 Trochoidal milling

applied, in addition to large axial depth of cut. In this way a large radial width of cut is avoided.

Slots wider than the cutting diameter of the tool can be machined, all with the same endmilling tool, usually an integral one. Since a small radial depth of cut is used, cutters with close pitch can be applied, leading to higher feed speed and cutting speed than with ordinary slot-milling applications.

A main drawback is that toolpath length is much higher compared to standard toolpaths such as zigzag because large tool movements are without engagement into the material. Moreover, in the case of sculptured surfaces, overlarge steps are produced on the surface, making very difficult the following semi-finishing operation. Therefore it is recommended for slotted shapes but not for free-form machining.

Currently all commercial computer-aided manufacturing (CAM) packages allow easy programming of this method.

2.9 Tools for Multitask Machining

Throughout the 2000s a new machine-tool concept called *multitasking machines* has been developed. This machine type is based on combining turning and milling operations in the same machine bed. Such solutions have been studied for over 20 years, mainly adapting turning centres equipped with a *C*-axis with mini-turrets for rotary tools. However, these machine tools were developed basically for turning operations, while milling operations were carried out with small tools with low power consumption (less than 1 kW). Moreover, the programming of these machines was a real challenge, as it was necessary to combine turning and milling cycles and it was necessary to program simultaneously four- or five-axis operations, with high collision probabilities.

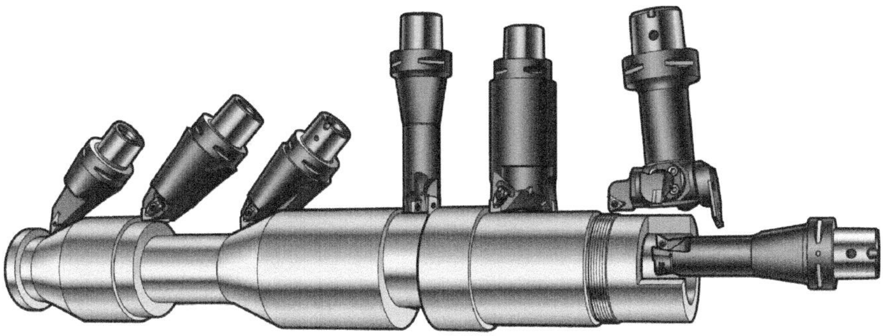

Figure 2.49 Tools specially designed for multitasking machining (Courtesy of Sandvik Coromant). The second from the right side is a mini-turret

These problems limited the development of these solutions until the 2000s, when a new series of machines were presented. The new multitasking machines are able to perform turning and milling operations without distinction and achieve the same power and accuracy of turning and machine centres. In short, multitasking machines include a spindle instead of a turret, in which a rotary tool (for milling, drilling, threading or hobbing) can be held, or a tuning head cab be locked. For more information on these machines the book about machine tools [21] is recommended.

The development of these solutions was also based on the use of latest-generation CNC and more reliable and powerful CAM programming systems. At present, multitasking machines are a reliable solution for the machining of complex parts, combining operations of turning, milling, drilling, boring, *etc.*, with the main advantage of making only one set-up, and consequently this fact allows a big reduction in lead times, increasing the machining accuracy.

In order to improve the results of multitasking operations, specific tools have been developed. In particular, a new design is based on two, three or four different turning inserts around the same holder. As the spindle of these machine tools can index its position, the different inserts can be oriented to combine different machining operations with the same holder, being known as *mini-turrets* (Figure 2.49).

2.10 Conclusions, the Future of Tools for Hard Machining

Machining is now in a particular "golden age", where a lot of time, money and effort has been invested to define the best tool for each application. Today for each application the objective of large or small manufacturers is to supply a much optimized tool, in all the related aspects discussed in this chapter. One of the most important aspects for the success of the new cutting tools is the *application guide*, because each application needs special recommendations, and in some cases they are contradictory to others.

The economical impact of cutting and machining is increasing, although the near to net shape technologies imply a reduction of the amount of material to be removed in each part. But the demand for elaborate parts and high-end products exceeds all expectations. Consequently the improvement of productivity, tool life and workpiece precision is a main goal for a lot of companies, taking into account respect for the environment as well.

Micromilling is going to be a growing technology where hard milling is going to be applied [22], with special attention to medical devices. In Figure 2.50 a test part used to study micromilling is presented. Tool fabrication is another important issue for the application of micromilling technology. For industrial applications, micropowder (0.3 μm particle size) sintered tungsten carbide is used, making two flute endmills of 100 μm in diameter, with an edge radius of 1–2 μm. In any case, the commercial offer is limited and there are no different geometries for different materials, being an important problem because most of the tools are designed for steel machining. Commercial tools have a well-defined geometry with small tolerances. Tolerance indicated in the catalogues for the sum of geometrical error plus runout error is of ±10 μm. However, real errors are usually smaller (±5 μm), but even in the best case, the tolerance with respect to size of the form to be machined is poor if compared to conventional high-speed machining mills. Tool wear (see Figure 2.51) is rapid and has a considerable effect on the process performance. It actually affects accuracy, roughness, and generation of burrs and vibrations.

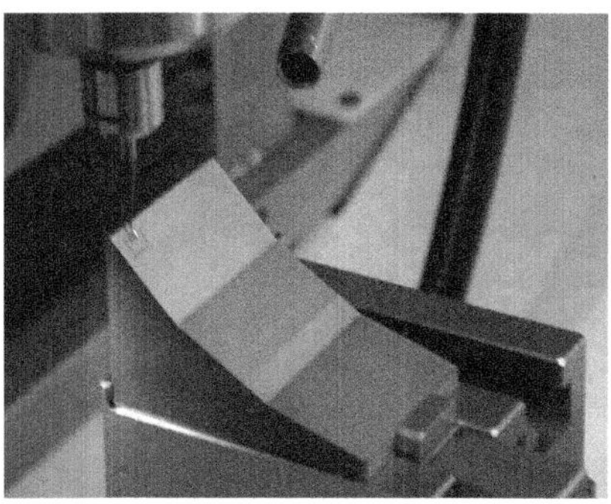

Figure 2.50 Micromilling of the test part made in 50 HRC hardened steel; two-tooth Ø 0.3 mm ball-endmill, 45,000 rpm, feed per tooth $f_z = 0.44$ μm/tooth, depth of cut $a_p = 8$ μm, radial penetration $a_e = 7.5$ μm

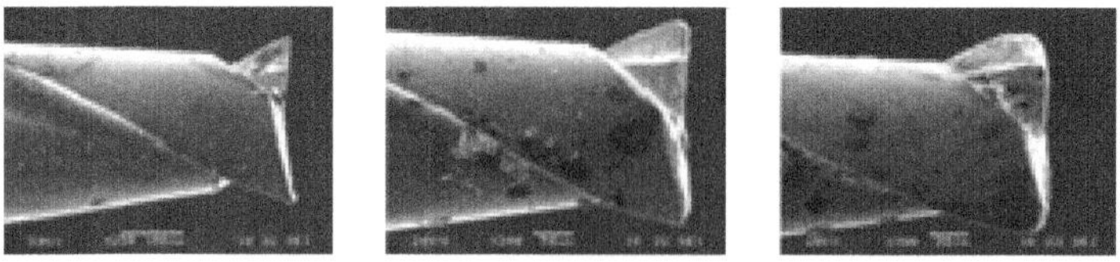

Figure 2.51 Tool wear evolution in micromilling (source: Tekniker)

On the other hand, materials with improved mechanical features are now in development, with more tensile strength and creep resistance. New alloys are usually very low-machinability alloys, asking for recommendation to be machined. Some examples are austempered ductile irons for car components and wind-energy gearboxes, gamma TiAl [23] for car components and aeronautical engines, high-silicon aluminium alloys, carbon-fibre-reinforced plastic composites [24], and others. Special tools will soon be on the market to solve the problems derived from the applications of these very difficult-to-cut materials.

Acknowledgements Special thanks are addressed to E. Sasia for his technical suggestions over the years, and Dr F. Campa, Dr L.G. Uriarte, Mrs A. Celaya, Mr D. Olvera, G. Urbicain and A. Fernández for their support. Financial support from the Spanish Ministry by project DPI 2007-60624, and Basque government by project SAIOTEK PROADI and ETORTEK Manufacturing 0,0 was received.

References

[1] HSS Smart Guide, 2010 available in www.hssforum.com
[2] ISO 513: 2004 Classification and application of hard cutting materials for metal removal with defined cutting edges – designation of the main groups and groups of application
[3] López de Lacalle LN, Lamikiz, A, Fernandes, M, Gutiérrez, A, Sánchez, J (2004) Turning of thick thermal spray coatings. J Thermal Spray Technol 10(2):249–254
[4] López de Lacalle LN, Lamikiz A, Muñoa J, Salgado MA, Sánchez JA (2006) Improving the high-speed finishing of forming tools for advanced high-strength steels (AHSS). Int J Adv Manuf Technol 29:1–2
[5] Schulz, H (1996) High-speed machining. Carl Hanser Verlag
[6] ISO 3685-1993 Tool life testing with single point turning tools
[7] ISO 8688-1-1989: Tool life testing in milling
[8] López de Lacalle LN, Sánchez JA, Lamikiz A, Celaya A (2004) Plasma assisted milling of heat-resistant superalloys. J Manuf Sci Eng 126(2):274–289
[9] Celaya A, López de Lacalle LN, Lamikiz A (2010) Assisted machining processes. In: Davim P (ed) Metal cutting: research advances. Nova Science Publishers, New York
[10] ISO 3002-1:1982, Basic quantities in cutting and grinding – Part 1: Geometry of the active part of cutting tools – General terms, reference systems, tool and working angles, chip breakers
[11] López de Lacalle LN, Lamikiz A, Sanchez JA, Salgado MA (2007) Toolpath selection based on the minimum deflection cutting forces in the programming of complex surfaces milling. Int J Mach Tools Manuf 47(2):388–400
[12] López de Lacalle LN, Lamikiz A (2008) Sculptured surface machining. In: Davim P (ed) Machining: fundamental and recent advances. Springer, London
[13] Lamikiz A, Lopez de Lacalle LN, Sanchez JA, Salgado MA (2004) Cutting force estimation in sculptured surface milling. Int J Mach Tools Manuf 44:1519–1526
[14] López de Lacalle LN, Lamikiz A, Sánchez JA, Arana JL (2002) Improving the surface finish in high-speed milling of stamping dies, J Mater Process Technol 123(2):292–302
[15] Lamikiz A, López de Lacalle LN, Sánchez JA, Bravo U (2005) Calculation of the specific cutting coefficient and geometrical aspects in sculptured surface machining. Mach Sci Technol 9(3):411–436
[16] Lazoglu I (2003) Sculpture surface machining: a generalized model of ball-end milling force system. Int J Mach Tool Manuf 43:453–462

[17] StarragHeckert (2007) 4/5 axis machining for complex workpieces.
[18] Salgado M, López de Lacalle LN, Lamikiz A, Muñoa M, Sánchez JA (2005) Evaluation of the stiffness chain on the deflection of end-mills under cutting forces. Int J Mach Tool Manuf 45:727–739
[19] Aoyama T, Inasaki I (2001) Performances of HSK tool interfaces under high rotational speed. CIRP Ann Manuf Technol 50(1):281–284
[20] Zelinski P (2010) Applying turn milling, mmsonline, available at www.mmsonline.com
[21] Lopez de Lacalle LN, Lamikiz A (2008) Machine tools for high performance machining. Springer-Verlag
[22] Uriarte L, Herrero A, Zatarain M, Santiso G , López de Lacalle LN, Lamikiz A, Albizuri A (2007) Error budget and stiffness chain assessment in a micromilling machine equipped with tools less than 0.3 mm in diameter. Precis Eng 31:1–12
[23] Beranoagirre A, López de Lacalle LN, (2009) Milling of gamma TiAl intermetallic alloys. AIP Conf Proc 1181:43–50
[24] López de Lacalle LN, Lamikiz A, Campa FJ, Fernández Valdivielso A, Etxeberria I (2009) Design and test of a multitooth tool for CFRP milling. J Compos Mater 43:3275–3290

Chapter 3
Mechanics of Cutting and Chip Formation

W. Grzesik

This chapter presents the basic knowledge on mechanics of the machining process in which the workpiece material hardened to 45–70 HRC hardness is machined with mixed ceramic or cubic boron nitride tools. Specific cutting characteristics, including cutting forces and cutting energy, and chip formation mechanisms are discussed in terms of process conditions. Additionally, currently developments in finite-element modelling are overviewed and some representative results are provided.

3.1 Mechanics of Hard Machining

3.1.1 Cutting Tools for Hard Machining

Hard machining is commonly defined as the process in which part pieces with hardness values over 45 HRC, but more typically in the 58–68 HRC range, are cut using specially prepared tools with geometrically defined cutting edges [1, 2]. The cutting-tool materials used are typically polycrystalline cubic boron nitride (PCBN), mixed (Al_2O_3-TiC) ceramics and sometimes cermets. The tooling choice will need to be matched to the application, desired production rates and the operating-cost goals. However, CBN is the most dominant choice for the more demanding applications of size and finish and particularly those components which have been transitioned from grinding. It is well known that in such extreme applications better tool performance can be achieved by using geometries with negative rake

W. Grzesik
Department of Manufacturing Engineering and Production Automation,
Opole University of Technology, P.O. Box 321, 45-271 Opole, Poland
e-mail: w.grzesik@po.opole.pl

angles but cutting-edge configuration and its preparation also play a dominant role in a machining success.

In general, the cutting-edge geometry is critical in hard machining because tools with superior edge strength are required to withstand the large tool stresses produced. As a result, tool cutting edges which are designed to cut hardened steels are often equipped with a chamfer (sometimes double chamfer) between the rake and clearance faces, termed by ISO the T-land and K-land respectively. The chamfer angle of –20° and the chamfer width of 0.1 (0.2) mm are typically selected. In order to protect the cutting edges from microchipping, additional hones are made as shown in Figure 3.1.

Because uniform edge micro-geometry along the corner radius causes excessive ploughing at the minor cutting edge (a very low edge radius to uncut chip thickness as shown in Figure 3.2 (a)) the PCBN tools with variable honed cutting edges are implemented [3, 4]. As can be seen in the CAD model of variable shaped cutting edge, the edge radius at points A, B and C decreases uniformly from the major to minor cutting edges, *i.e.* $r_{\beta A} > r_{\beta B} > r_{\beta C}$. This causes a constant uncut chip thickness to edge radius ratio to be maintained along the active part of the corner.

Recently, wiper inserts, which combine the high feed capability and high-quality surface finish produced by large round inserts, have gained popularity in hard-machining applications. As shown in Figure 3.3 (a), multi-radii tool corner contains a small smoothing part of the radius of r_{bo} which is parallel to the feed direction. Additionally, Figure 3.3 (b) shows a more universal design of a wiper corner with both left- and right-handed wiper segments. By the application of such a solid wiper insert (the Crossbill™ by Seco Tools [6]), it is possible to machine a perfect radius with no deviation from a normal radius along with axial turning, which utilizes the full wiper effect.

Figure 3.4 shows a unique new design of solid CBN inserts with a straight part of the cutting edge blending into a wiper. As a result, the smaller approach angle κ_r reduces the depth of cut and provides constant chip thickness, so higher feed rates up to 0.4 mm/rev can be applied. Figure 3.5 illustrates the micro-geometry of a few of the inserts at a magnification of 50 times obtained by using field emission scanning electron microscopy.

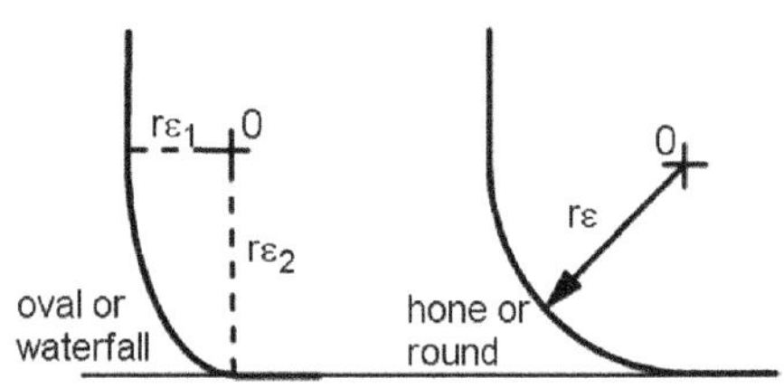

Figure 3.1 Profiles of honed cutting edges for hard-machining applications [3]

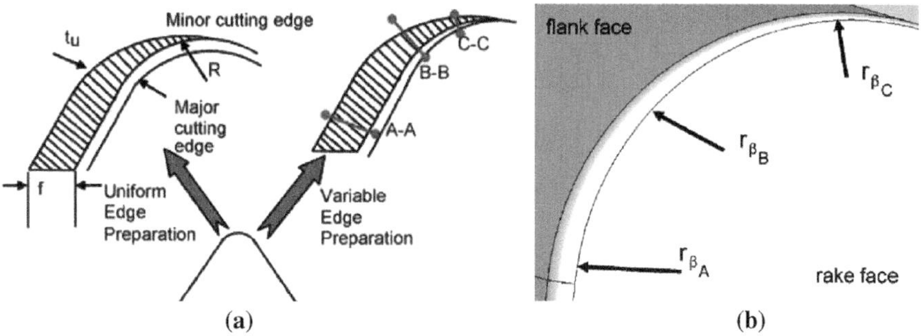

Figure 3.2 Comparison between uniform and variable micro-geometry of the corner (a) and CAD model of the variable-hone edge design (b) [3, 4]

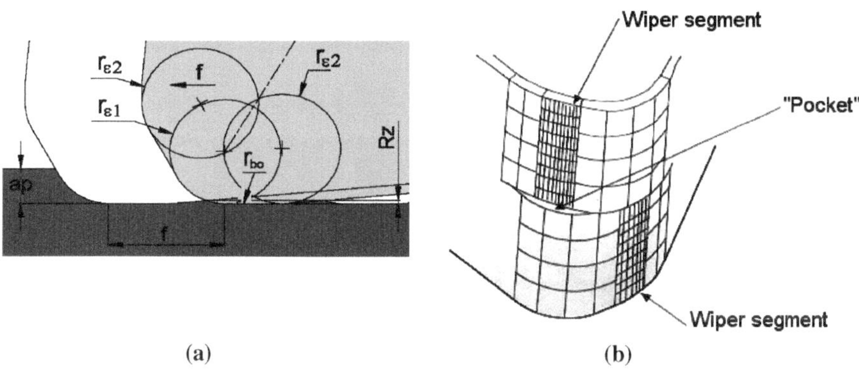

Figure 3.3 CBN wiper inserts [5, 6]: (a) one-handed design (*f*: feed; a_p: depth of cut; $r_{\varepsilon 1}$ and $r_{\varepsilon 2}$: radii of wiper curvature; r_{bo}: radius of smoothing part; R_z: valley-to-peak height), and (b) two-handed design

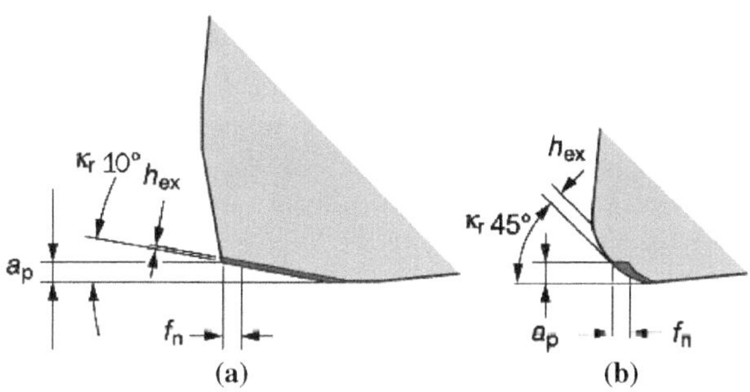

Figure 3.4 Xcel geometry and comparison of chip thicknesses for modified wiper (a) and conventional corner (b) [6]

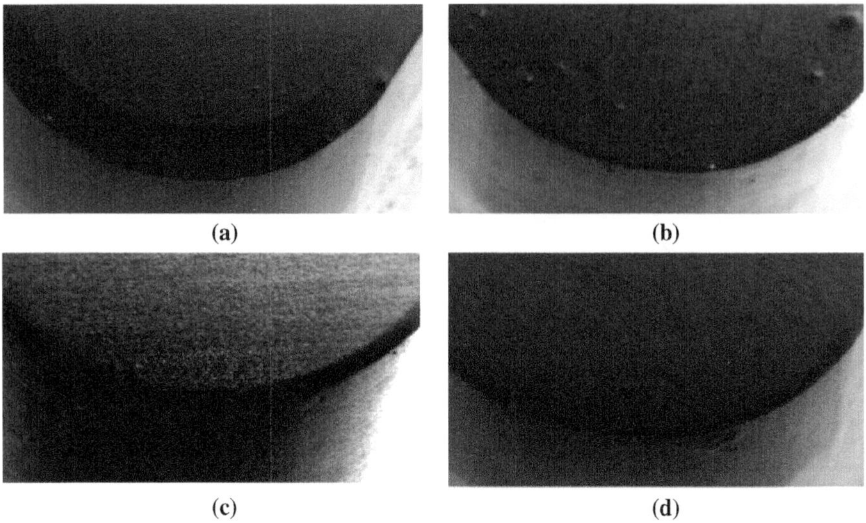

Figure 3.5 PCBN inserts with different micro-geometries of cutting edges: (a) chamfer insert ($20° \times 0.1$ mm), (b) uniform hone ($r_\beta = 50$ µm), (c) waterfall-hone insert ($r_{\beta 1} = 30$ µm, $r_{\beta 2} = 60$ µm), and (d) variable-hone insert ($r_{\beta A} = 50$ µm, $r_{\beta B} = 10$ µm) [4]

3.1.2 Mechanical Models of Hard Machining

Hard machining is a specific process performed under unique technological and thermo-mechanical conditions and, as expected, the cutting-process mechanisms (chip formation, heat generation, tool wear) differ substantially from those observed in machining soft materials. As described in [1, 7, 8], hard machining is performed also as a dry high-speed machining process. In particular, while small depths of cut (0.05–0.3 mm) and feed rates (0.05–0.2 mm/rev) are used, both small undeformed chip thickness (UCT) and the ratio of the UCT to the radius of the cutting edge are obtained in such processes. These geometrical relationships lead to an effective rake angle of –60 to –80° and, as a result, extremely high pressure is generated to remove material in the vicinity of the cutting edge. Moreover, a large corner radius causes the components of the resultant cutting force to increase, along with extremely high thermal stresses.

The principal model of the orthogonal cutting process with a chamfered tool proposed by Ren and Altintas [9] is shown in Figure 3.6. It should be primarily noted that this model incorporates a small dead metal zone OBH created by the chamfer in contrast to the obviously known models with both primary and secondary plastic deformation zones. As a result, the plastic flow under the chamfer edge follows in the direction opposite to the tool motion in order to avoid the negative energy dissipation in the cutting zone. The proposed model utilizing appropriate slip-line solution is able to generate the optimal chamfer angle and cutting speed (so it will be useful in modelling of high-speed machin-

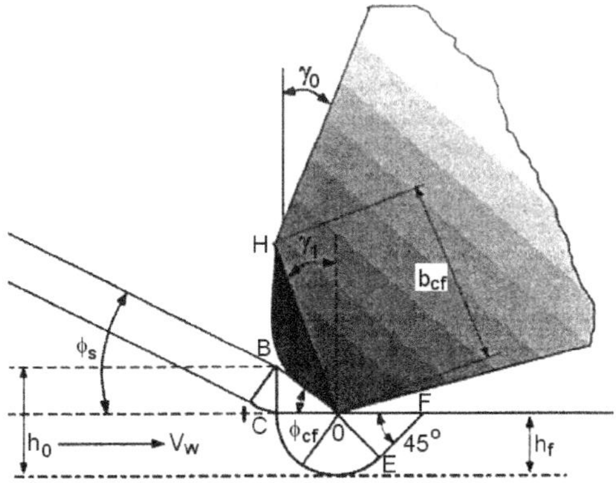

Figure 3.6 Model of orthogonal machining with chamfered edge tools [9]

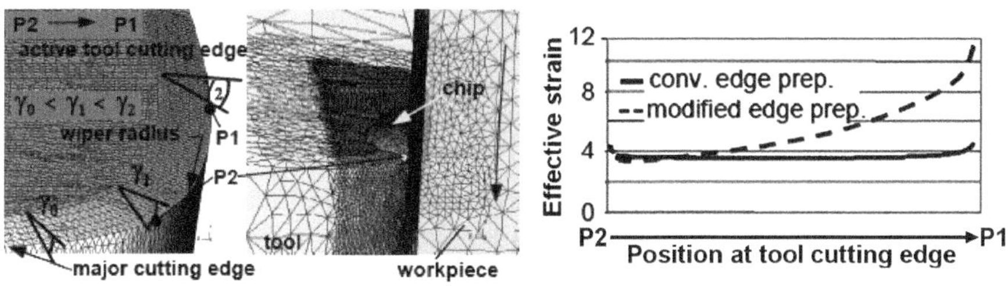

(a) (b)

Figure 3.7 Wiper insert with local variable rake angles (a) and distribution of effective strain (b) [10]

ing of hardened materials) and guarantee lowest tool wear and relatively low cutting forces.

As reported in Section 3.1.1, cutting edges of PCBN tools for hard turning typically have a constant value of the negative rake angle on the major first face. In the case shown in Figure 3.7 (a), wiper inserts were prepared in such a way that the inter-linked tool edge combines micro-parts of round and rhombic tool inserts and provides variable chamfer width and continuous variation of the chamfer angle up to $\gamma_2 = -60°$ [10]. In consequence, the local chip thickness approaches zero at the position of the wiper radius, where the peripheral land corresponds to the bluntest chamfer. Moreover, as shown in Figure 3.7 (b), the effective plastic strain increases within points P1 and P2, and, as a result, the cutting-edge bluntness is correlated with the degree of plastic deformation of the machined surface.

3.1.3 Cutting Forces

Specific cutting conditions of the hard-machining process and complex microgeometrical features of cutting tools used should result in different behaviour of the process including process mechanics. It suggests that values of force components

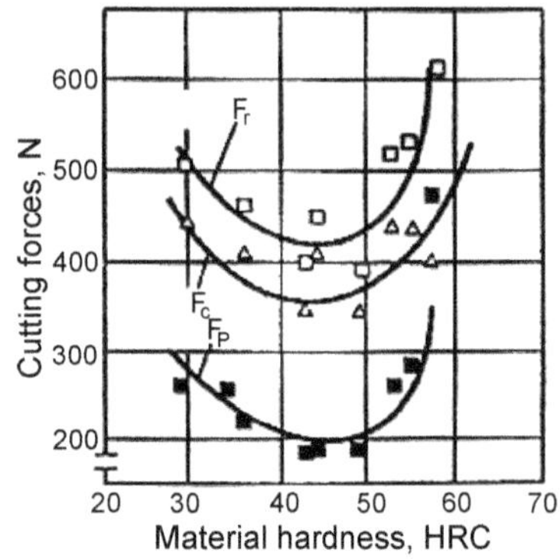

Figure 3.8 Influence of steel hardness on cutting forces (v_c = 90 m/min, f = 0.15 mm/rev, a_p = 0.9 mm) [2]

and relationships between them should differ from those obtained in machining of soft materials. First, it can be observed in Figure 3.8 that cutting forces increase drastically when machining materials with hardness higher than about 45 HRC (this value is often referenced as a lower limit of hard-part machining).

In particular, larger negative rake angle and tool corner radius influencing the passive force F_p increases remarkably and this effect means that an absolutely stable and rigid process has to be provided. This requirement has to be especially kept when using super-hard tools with multi-radii smoothing geometry (so-called wiper tools).

It is shown in Figure 3.9 that for micro-alloyed 30MnVS6 and quenched–tempered (Q-T) AISI 1045 and AISI 5140 steels, all hardened to about 300 HBN, the cutting and trust force variations with an increase of cutting speed from 10–250 m/min revealed characteristic peaks at the cutting speed of about 50 m/min [11] independent of steel grades. It should also be noted that the effect of thermal treatment (Q-T) is more pronounced than micro-alloying at the same cutting conditions. Similar curvature was obtained for the force-feed function. According to the experimental data provided, the cutting speed of about 200 m/min seems to be optimal for appropriate machining operations.

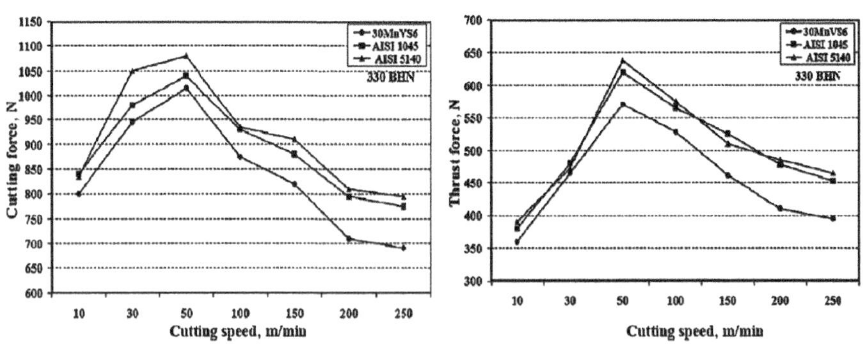

(a) (b)

Figure 3.9 Influence of cutting speed on cutting (a) and thrust (b) forces for three hardened steels of 300 BHN using TiN-coated carbide tools, f = 0.11 mm/rev [11]

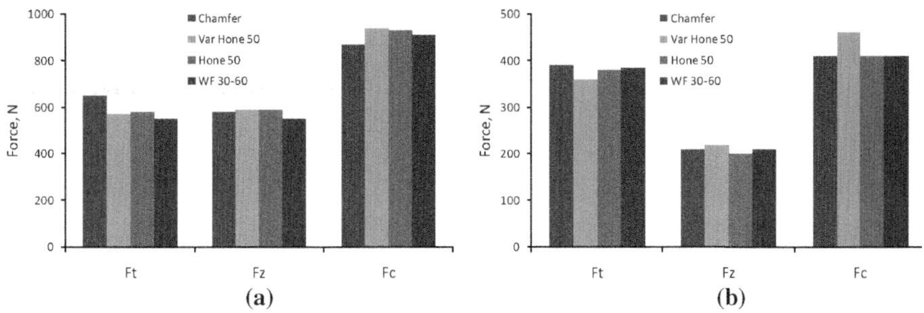

Figure 3.10 Measured cutting forces in turning of hardened AISI 4340 steel: (a) $v_c = 125$ m/min, $f = 0.15$ mm/rev, $a_p = 1$ mm, and (b) $v_c = 175$ m/min, $f = 0.15$ mm/rev, $a_p = 0.5$ mm [3]

The average values of measured components of the resultant cutting force for different edge micro-geometry shown in Figures 3.1 and 3.2 and two cutting speeds of 125 and 175 m/min are presented in Figure 3.10. For the cutting speed of $v_c = 125$ m/min, waterfall hone edge with 30:60-μm edge radii yielded the lowest radial force (F_t) followed by variable-hone 50-μm edge radius. On the other hand, the latter design produced the highest cutting force (F_c). Similarly, at the higher tested cutting speed of 175 m/min (Figure 3.10 (b)), PCBN tools with variable-hone cutting edges (VarHone50) produce higher cutting forces but measured radial forces decrease in comparison to other cutting-edge geometries. In addition, it is revealed [10] that for modified wiper geometry (Figure 3.7 (a)) the angle between the resultant cutting force and the resultant passive force is significantly reduced when the chamfer angle varies from $\gamma_1 = -10°$ to $\gamma_2 = -60°$.

As presented in Figure 3.11, the cutting-edge preparation has a significant effect on the cutting forces generated in high-speed milling of H13 tool steel, heat treated to hardness of 55 HRC, using ball-nose endmills with brazed CBN inserts. In this case, the highest values of the resultant cutting force (F_r) were obtained for honed edges for which the large negative effective rake angle results in the intensification of the ploughing effect, which visibly increases the friction force component. On the other hand, the lowest cutting force was recorded for chamfered cutting edges when using higher feed rates of 0.1 mm/tooth. This is because increasing the feed for chamfered cutting edges results in a more stable cutting process.

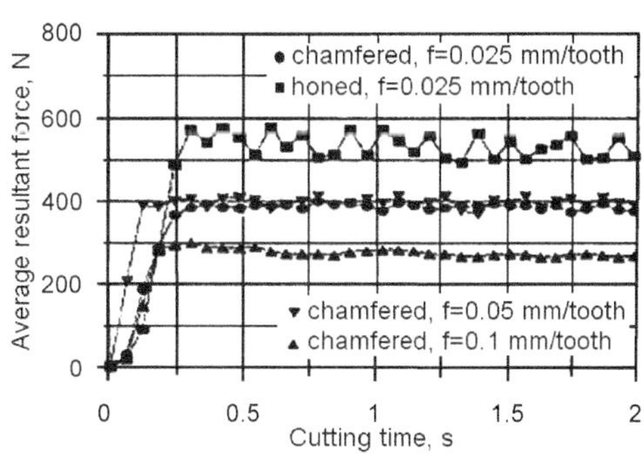

Figure 3.11 Effect of feed and edge preparation on cutting forces in milling of hardened AISI H13 tool steel [12]

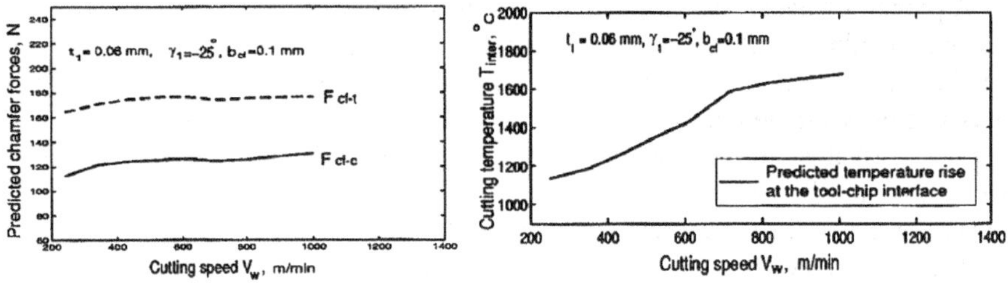

(a) (b)

Figure 3.12 Influence of cutting speed on chamfer forces and interface temperature [9]

Figure 3.12 illustrates how cutting speed influences both the chamfer force ($F_{cf\text{-}t}$ and $F_{cf\text{-}c}$ in the feed and cutting directions, respectively) and corresponding cutting temperature when using CBN chamfered tools with 0.1-mm chamfer width and –25° chamfer angle. In this study, three cutting speeds of 240, 600 and 1000 m/min were selected. As can be seen in Figure 3.12 (a), the chamfer forces seem to be independent of cutting speed rise because the temperature rise in the primary deformation zone is rather small, not exceeding 50 °C. On the other hand, the temperature in the chip–rake-face contact zone rises significantly up to above 1500 °C at cutting speed of 600 m/min. A practical recommendation is that CBN tools can operate effectively up to 600 m/min because they do not exhibit significant crater (diffusive) wear and micro-cracks.

For the given material couple Al_2O_3/TiC–AISI D2 of about 60 HRC, the dependence of the cutting force on depth of cut for conventional and wiper insert configurations and three machining times (5, 10 and 15 min) are shown in Figure 3.13. In particular, the cutting forces recorded for wiper tools (Figure 3.13b) change almost linearly but for conventional tools the F_c–a_p functions have visible maximum points at depths of cut in the vicinity of $a_p = 0.45$ mm. Probably, in the latter case tool wear causes the geometry of the cutting edge to become less negative, in contrast to wiper tools for which tool wear is substantially lower [14] (for wiper tools the increase of cross-sectional area of cut is the predominant effect).

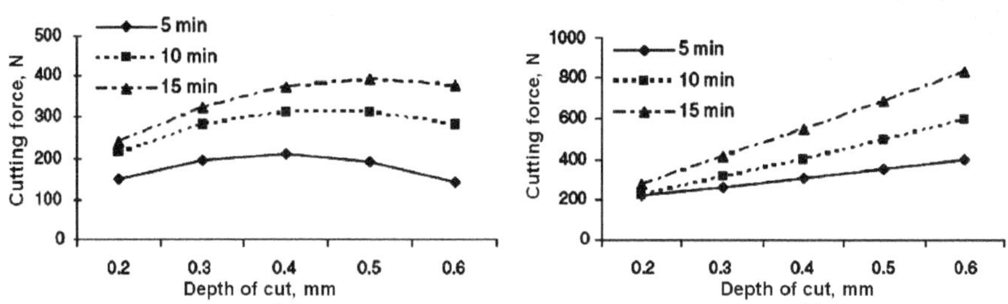

(a) (b)

Figure 3.13 Influence of depth of cut on cutting force for $f = 0.1$ mm/rev, $v_c = 80$ m/min for conventional (a) and wiper (b) inserts in machining of AISI D2 cold-work tool steel with Al_2O_3-TiC ceramic (CC650 grade) tools [13]

Arsecularatne *et al.* [15] have investigated variations of cutting (F_c), feed (F_f) and radial (F_r) forces with a machining time of a few minutes in turning hardened AISI D2 steel of 62 HRC with PCBN tools. The most feasible feeds and speeds fell in the ranges of 0.08–0.20 mm/rev (0.14 mm/rev for finishing and 0.2 mm/rev for roughing operations) and 70–120 m/min, respectively. It was found experimentally that the cutting force shows a little increase with machining time and F_c is the largest force, while F_f is the smallest force. Generation of a larger F_r than F_f is common in hard turning.

3.1.4 Cutting Energy

The specific cutting energy (SCE) is one of important process indicators which quantify the level of energy consumption and influence the industrial applications of a given machining operation. Figure 3.14 shows the SCE for orthogonal turning of AISI 52100 bearing steel (60 HRC) using a low-CBN-content (70 %) tool for various tool and cutting geometries, *i.e.* for variable rake angle and UCT. According to the data presented in Figure 3.14, the SCE for this hard-machining operation ranges from 6 to 11 GPa, which is substantially higher than for machining conventional, softer steels. Not surprisingly the highest SCE was obtained for tools with very large negative rake angle ($\gamma_0=-27°$) and small UCT of 15.4 μm. Moreover, the SCE decreases with cutting speed rise but predominantly for tools with higher negative rake angles.

According to the common practice of hard machining, the high SCE required to machine hardened steels lowers the critical depth (or width) of cut for which regenerative chatter occurs. Thus, to maintain high stability at reasonable metal-cutting rates, the stiffness and damping of the machine-tool system must correspondingly be high.

At very low feeds, the UCT is considerably less than the radius at the tool tip and the effective rake angle becomes largely negative. The material ahead of the rake face is in an intensely compressive stress state. In such a case, a large volume of material becomes fully plastic before a very thin chip is formed. This causes an exponential increase in specific energy with smaller UCT values, as shown in Figure 3.15. By analogy, hard turning seems to be similar to the grinding process.

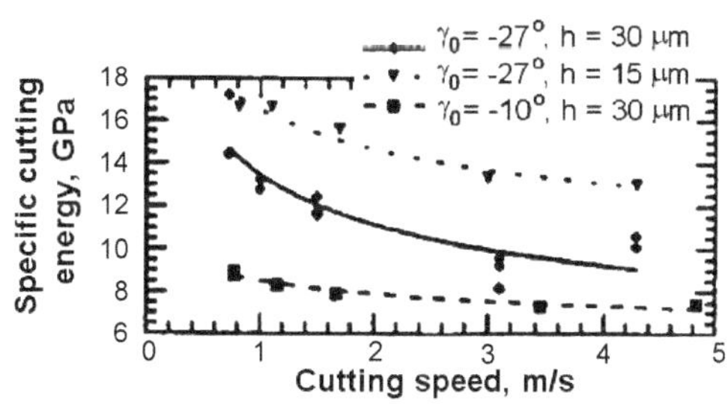

Figure 3.14 SCE *vs.* cutting speed for orthogonal turning of hardened bearing steel [16]

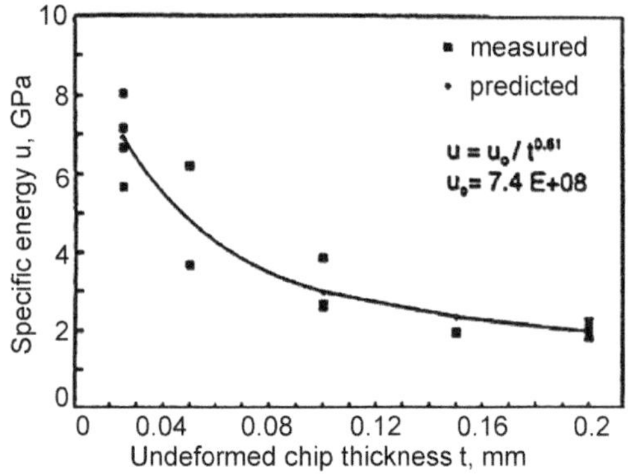

Figure 3.15 SCE *vs*. UCT for case-hardened steel [17]

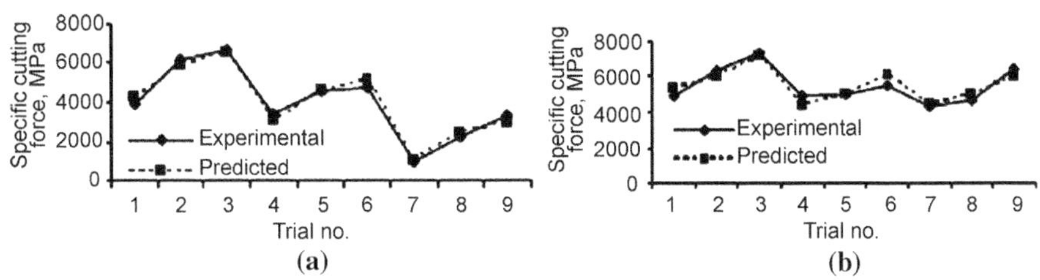

Figure 3.16 Comparison of specific cutting forces in turning of hardened cold tool steel using conventional (a) and wiper (b) mixed ceramic (CC650 grade) inserts [13]

However, the main difference (see Section 3.2.1) is that fracture is the root cause of chip formation in hard machining.

Figure 3.16 shows how specific cutting force (energy) changes for different CC650 ceramic-insert configurations (conventional, wiper general and wiper hard-part turning) when machining high-chromium AISI D2 cold-work steel of average hardness 59/61 HRC. In this investigation the trial time was varied from 5–15 min and the depth of cut was changed from 0.2 to 0.6 mm. The cutting speed of 80 m/min and feed of 0.1 mm were kept constant. It is very interesting in Figure 3.16 that SCE varies from 1035 to 6670 MPa and from 4345 to 7310 MPa for conventional and wiper inserts respectively. The lowest values of SCE were obtained for high depth of cut of 0.6 mm and 5 min cutting tests, and this trend qualitatively agrees with data gathered in Figures 3.14 and 3.16.

3.1.5 *Influence of Supply of Minimum Quantity of Lubricant on Mechanical Behaviour of Hard Machining*

Recently, the applicability of the minimum quantity of lubricant (MQL) technique has been tested in high-speed hard-machining operations, such as turning or milling, not only in order to alleviate the pollution problems but predominantly to

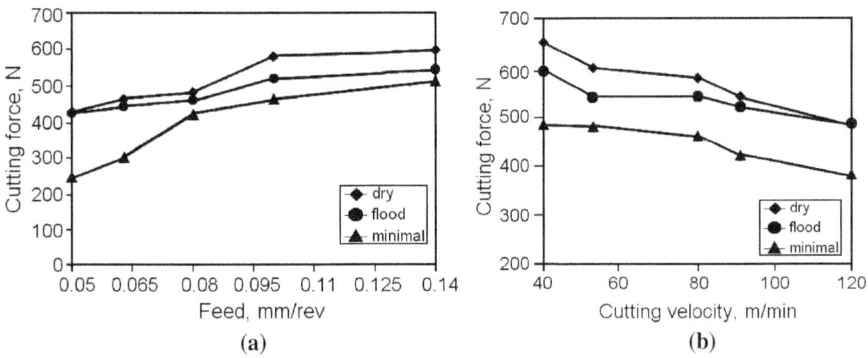

Figure 3.17 Cutting forces *vs.* feed at $v_c = 80$ m/min, $a_p = 1.25$ mm (a) and cutting speed at $f = 0.10$ mm/rev, $a_p = 1.25$ mm (b) for dry, wet and MQL machining conditions [18]

reduce tool wear and improve surface integrity, especially to minimize white layer formation and induce the desired residual stress profile in the subsurface layer [17–19]. For instance, in turning of a through-hardened AISI 4340 steel of 46 HRC hardness with coated carbide inserts, a thin pulsed jet of special cutting fluid at velocity of 100 m/s and pressure of 20 MPa was supplied at the immediate cutting zone at a low rate of 2 ml/h [18].

As a result, the performance of hard turning with minimal fluid application in terms of cutting forces was changed in comparison to dry and wet conditions, as shown in Figure 3.17. It can be easily observed that the cutting force is substantially lower in a range of feed rates and cutting speeds when compared to dry turning and conventional wet turning. Presumably, minute capillaries are formed at the tool–chip interface and they reduce the friction contribution to the cutting force. As revealed, reduced contact length and compression chip ratio (thinner and shorter chips were produced) are indicated of gentle frictional conditions at the tool–chip interface.

When the milling of ASSAB F3 hardened tool steel of hardness 51 HRC with TiAlN coated carbide ball-endmills was assisted by a special fluid injection system, the progress of cutting force components during the process was as reflected

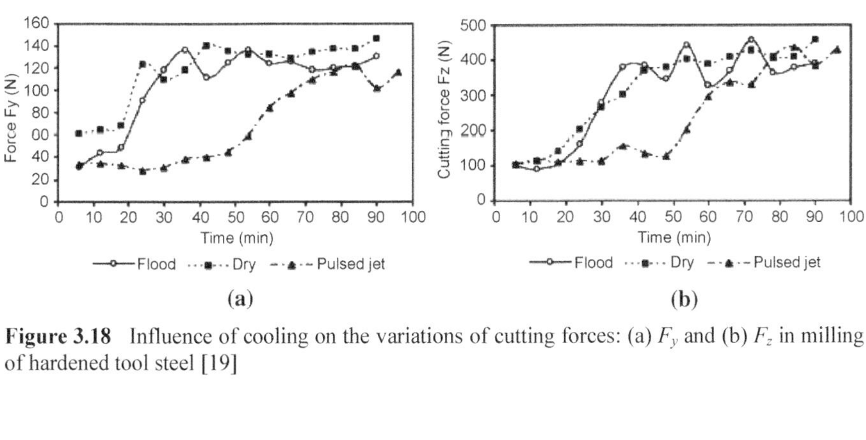

Figure 3.18 Influence of cooling on the variations of cutting forces: (a) F_y and (b) F_z in milling of hardened tool steel [19]

in Figure 3.18. Considering the whole period of cutting, it can be seen that, in general, cutting forces in pulsed-jet mode are lower than in flood and dry milling. However, the values of cutting forces are almost the same when tool failure criterion of 0.35 mm was achieved. Similarly as for turning, due to better lubrication, the wear curve does not express a rapid increase of flank wear at the beginning of cutting, *i.e.* in the running-in period.

3.1.6 Finite-element Modelling of Mechanical Loads

As reported in many papers, for example those recently published [3, 20], which overview the state-of-the-art of numerical modelling of hard machining, finite-element modelling (FEM) can be extremely useful for understanding physical phenomena and mechanisms which govern chip formation in machining of hardened steel materials. In this specific problem care has to be taken to provide appropriate flow-stress data and friction model, and to select the predominant mechanism of chip formation, *i.e.* surface shear-cracking (SSC, originally termed shear-crack hypothesis) or catastrophic thermoplastic instability (CTI). According to Kountanya *et al.* [20], the SSC mechanism predominates in a majority of hard-machining conditions. In the past, 2D FEM of temperature and forces was developed [21, 22] but recently 3D FEM has been applied for analysing a hard-turning process [3, 23]. It is assumed that a hard-turning process is a typical high-speed dynamic event with large nonlinear material deformation at high strain rates and temperatures, complex contacts and material failure. The explicit simulation method seems to be suitable for hard-machining processes due to small increments (small feed and depth of cut).

Figure 3.19 shows the comparison of measured and simulated cutting forces for four different micro-geometries of PCBN inserts keeping $v_c = 300$ m/min, $f = 0.15$ mm/rev and $a_p = 1$ mm. It is clear that the FEM procedure using the John-

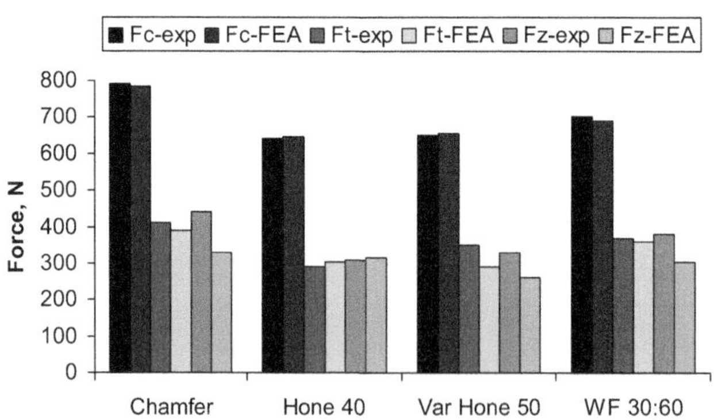

Figure 3.19 Comparison of measured and simulated forces for different micro-geometry of PCBN tools [3]

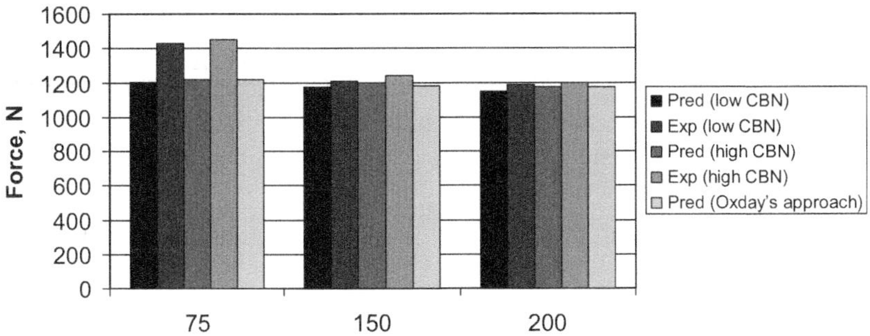

Figure 3.20 Comparison of predicted and measured values of cutting force for different CBN-content tools and variable cutting speed [26]

son–Cook (JC) material constitutive model, applied by Özel [4], which adopts pressure-dependent shear friction, gives acceptable agreement between experiment and simulation. Similarly, in the FEM simulations with the coupled flow-stress model performed by Kountanaya *et al.* [20], the maximum errors in predicting cutting F_C and thrust F_T forces were 20 % and 9 %. Chen *et al.* [24] have established that the resultant cutting force for hone edges is lower but such tools can be employed for hard turning when the tensile principal stresses (TPS) acting on the tool can be reduced to a low magnitude. On the other hand, chamfer edges are more robust because their use allows the TPS to be reduced by about 25 % [25], which significantly reduces the probability of catastrophic failure.

Huang and Liang [26] have modified the JC model by considering Oxley's machining predictive theory, in particular to represent the workpiece material property as a function of strain, strain rate and temperature. This model was validated experimentally for AISI H13 tool steel of 52 HRC hardness and both low-CBN-content and high-CBN-content tools.

The effect of different CBN content tools and variable cutting speed on predicted values of cutting force in turning of AISI H13 tool steel (52 HRC) is presented in Figure 3.20. It is observed that both the cutting (tangential) and thrust forces are higher when generated by the high-CBN-content tools. As discussed, the differences in force prediction can result from variation in material hardness and the effective rake angle assumed.

3.2 Chip Formation in Hard Machining

3.2.1 Criteria for Crack Initiation and Propagation

One of the characteristic phenomena occurring in the machining of hardened steels with geometrically defined cutting tools is the formation of saw-tooth chips. Catastrophic failure causing saw-tooth chip formation in the primary shear zone is usually attributed to either cyclic crack initiation and propagation or to the occurrence of a thermoplastic instability [27, 28]. Section 3.1.6 described these mechanisms as SSC and CTI [20]. The periodic crack theory assumes that periodic shear cracks first develop near the surface of the work and proceed downward along the shear plane toward the tool tip. Following crack formation, bands of concentrated shear may or may not then develop [29].

Figure 3.21 shows the cyclic mechanism of the formation of chip segments due to crack initiation (numbered successively 1 and 2 close to the machined surface) when the UCT is higher than 0.02 mm. This is because continuous chips are formed for very small UCT less than $h < 0.02$ mm.

Associated stress and temperature fields developing in the vicinity of the cutting edge are shown in Figure 3.21. It can be seen in Figure 3.21 that in machining of hardened 100Cr6 bearing steel, the direct normal stresses σ_{VB} of the magnitude of approximately 4000 MPa result in high mechanical and thermal stresses extended to the machined surface of the workpiece. Thermal stresses originate mainly in the intensive friction between flank wear land and the workpiece, which for the friction coefficient of 0.2–0.3 causes high tangential stress [31]. The temperature field in the workpiece due to friction, predicted by assuming a semi-infinite moving body with an adiabatic surface and the heat partition to the workpiece of 80 %, is as shown in Figure 3.21 (b).

In cases for which the chip speed is high enough, the temperature near the machined surface may reach the γ–α transition temperature, martensite produced by friction development can form the so-called white layer observed in chip micrographs.

Elbestawi *et al.* [17] developed the chip formation model, in which it starts with initiation of a crack at the free surface of the workpiece, by considering the direction of crack initiation using the surface layer energy/strain energy density criterion. It is assumed that the machined surface is rough and notched, and it locally has very small elliptical notches, as shown in Figure 3.22 (a). The compressive cutting pressure p acts at a certain angle Φ from the major axis of the elliptical notch. According to this hypothesis, the crack at the free surface will initiate at a critical angle Φ_{cr} when surface layer energy reaches its maximum value at the minimum applied pressure (Figure 3.23 (b)). For AISI 1550 case-hardened steel (60 HRC) machined with mixed ceramic tools at $v_c = 66$–120 m/min, $f = 0.025$–

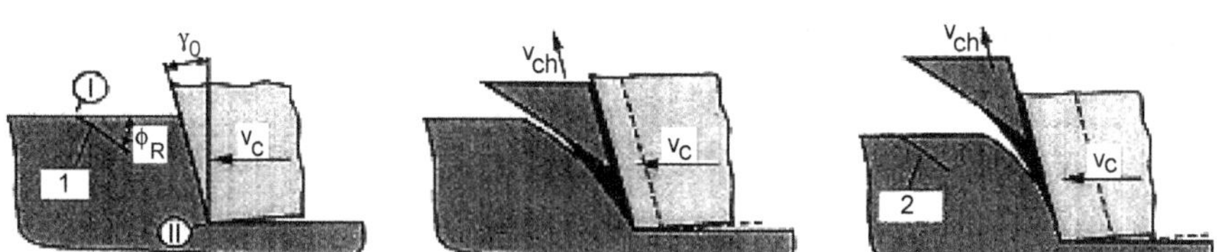

(a) (b) (c)

Figure 3.21 Chip formation mechanisms for hardened 100Cr6 (60–62 HRC) steel and UCT of 0.05 mm (when $h > 0.02$ mm) when using PCBN tool: (a) first stage, (b) intermediate stage, and (c) second stage [30]

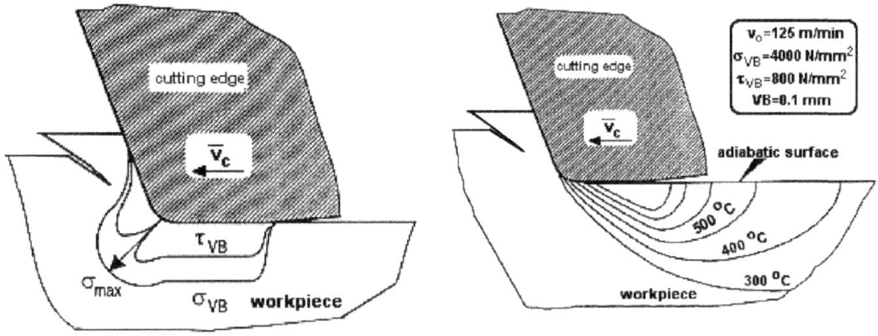

Figure 3.22 Associated distributions of mechanical stresses (a) and temperature (b) before crack propagation for machining 100Cr6 (63 HRC) steel. Cutting conditions: cutting speed of 125 m/min and tool wear VB = 0.1 mm) [31]

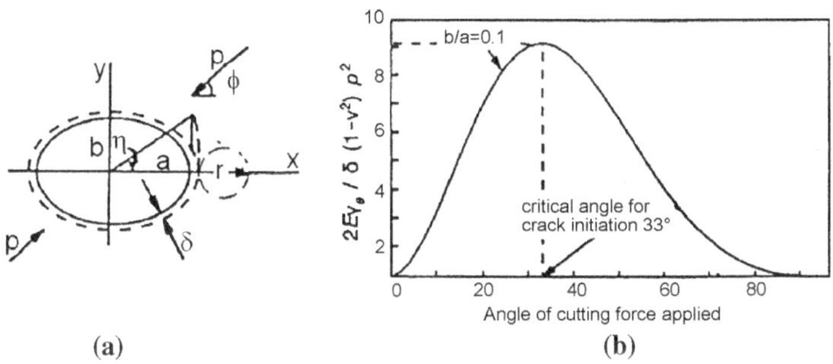

Figure 3.23 Unit elliptical notch under compression (a) and variation of maximum surface layer energy with loading angle Φ [17]

0.2 mm/rev and a_p = 0.5–2 mm, the maximum surface energy was obtained for the crack initiation angle equal to 33°. For this case of hard machining, the relevant angle of crack propagation was found to be equal to about Θ = 55°. The value of the fracture angle higher than 45° suggests that in hard turning the chip formation mechanism differs from the pure shear deformation process only.

3.2.2 *Criteria for Shear Instability*

According to the adiabatic shear theory [1], the root cause of saw-tooth (segmental) chip formation is a CTI for which the decrease in flow stress due to thermal softening exceeds the associated strain hardening. Materials sensitive to the formation of localized shear chips have either poor thermal properties (titanium alloys, nickel-based superalloys) or limited ductility, such as hardened alloy steels.

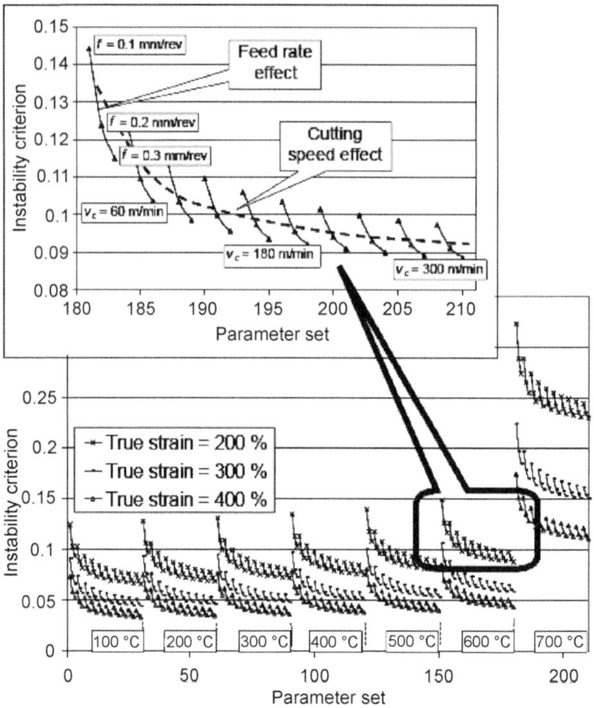

Figure 3.24 Shear instability criterion for hardened 100Cr6 steel [22]

The first criterion for shear instability was formulated by Recht (see Chapter 7 in [1] or *Trans. ASME, J. Appl. Mech.*, Vol. 31, 1964, 189–193). It assumes that the slope of the strain-hardening curve becomes negative when an increase in shear strain is associated with a decrease in shear stress [22]. This means that shear localization occurs when values of the criterion R given by Equation 3.2 will be between 0 and 1. If this ratio is equal to 1, a catastrophic slip will dominate.

$$R = \frac{\dfrac{\partial \overline{\tau}}{\partial \overline{\gamma}}}{\dfrac{\partial \overline{\tau}}{\partial T}\dfrac{dT}{d\overline{\gamma}}}, \text{and } 0 \leq R \leq 1 \tag{3.1}$$

where $\overline{\tau}$ is shear stress and $\overline{\gamma}$ is shear strain.

Another criterion, proposed by Semiatin and Rao, uses the strain-rate-dependent β parameter, for which shear localization predominates when it is greater than 5 [22].

Figure 3.24 shows the template of instability criterion for the 210-parameter set $(10v_c \times 7T \times 3f \times 3\overline{\gamma})$ when T= 100–700 °C, $\overline{\gamma} = 2-5$, v_c=30–300 m/min and f=0.1–0.3 mm/rev. The shear strain was determined based on standard hot compression tests for temperature varying from 20 to 700 °C (relevant true stress–true strain curves are presented in the lower diagram in Figure 3.24). It should be noted that for the material and cutting conditions used, the value of R criterion is always between 0 and 1 but is not higher than 0.15. Moreover, the increases in all variables (cutting speed and feed rate) result in the decrease of R value.

3.2.3 Mechanisms of Chip Formation

The nature of chip formation in hard machining is quite different than in more conventional machining. After reviewing the mechanics of this unique type of chip formation some mechanisms which govern it will be explained and discussed. The first, simple periodic-cracking based model for producing saw-tooth chips shown in Figure 3.25 (a), was proposed by Shaw [29]. At point C (case (i)) material in the free surface begins to rise and assumes a direction CD parallel to the resultant cutting force F_r. A shear crack initiates at point D and develops downard along shear plane DO toward the rake face (case (ii)). When the tool moves, the chip slides along the cracked surface until the next crack forms at point D′ (case (iii)). Initially the crack, called a *gross crack* (GC), will be continuous across the width of the chip for sufficiently brittle materials, but for less brittle materials at higher cutting speeds, it may become discontinuous as the crack proceeds towards the tool tip. Such disconnected localized cracks will be termed *microcracks* (MC). The distance between one segment that slides relative to its neighbour during one cycle depends upon the distance p (C′D′) between cracks on the machined surface. When the chip pitch p_c (Figure 3.28) is higher than p (chip compression ratio $k_h < 1$ and shear angle greater than 45°), this is a result of the compressive stresses on the material in the MC zone. The thinning of the MC region is usually the case when hard steel is turned with a negative rake tool. On the other hand, material in the GC region is carried along with the MC material, resulting in k_h for the entire chip being greater than one. There are then two regions as the chip proceeds up the tool face – the material between GCs sliding outward and deformation in the MC region resulting in bending downward and running along the tool face, the concentrated shear bands.

Based on experimental observation, Shaw and Vyas [29] reported that in face milling of case carburized AISI 8620 steel (61 HRC) with PCBN tools at $v_c = 150$ m/min, $f = 0.13$ and 0.25 mm/rev and $a_p = 0.13$ and 0.25 mm the chip formation is of a cyclic saw-tooth type.

In another mechanical model of chip segmentation, segmented chips are produced by catastrophic strain localization occurring above some critical cutting speed. As shown in Figure 3.25 (b), catastrophic shear is initiated along the line B′D′ (inclined at an angle Θ relative to the workpiece surface) when the tool tip is at point B′. As the tool moves from B′ to A, the stress required for further deformation along BD decreases due to the thermal softening and the tool experiences rapid unloading along BE. At the same time, the loads increase along AB as the material in triangular region B′AB is indented and sheared.

Because the shear zone is nearly adiabatic and the workpiece temperature ahead remains essentially at ambient, the formation of the next segment is determined highly by the stresses applied to the workpiece along AB. In this model the

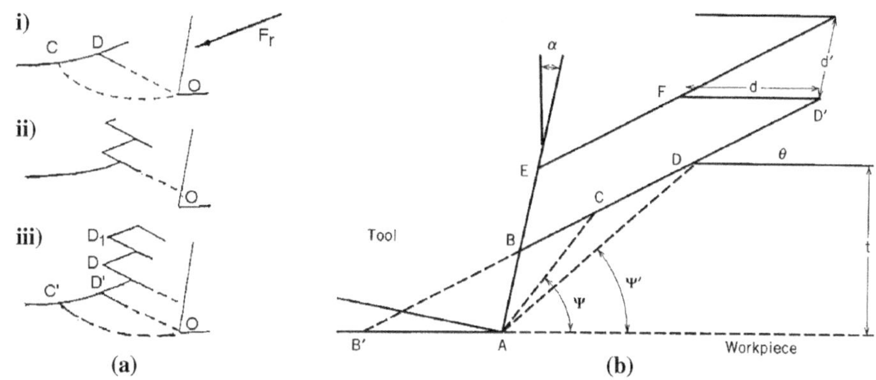

Figure 3.25 Mechanism of saw-tooth chip formation after Shaw [29] (a), and simple model of segment formation after Davies *et al.* [16] (b)

segment spacing is a linear function of the UCT but increases asymptotically with cutting speed, depth of cut and rake angle. For instance, for the orthogonal machining of a through-hardened AISI 52100 bearing steel of 50–65 HRC with PCBN tools, the onset of chip segmentation due to adiabatic shear was observed at relatively low cutting speeds (below 1 m/s) [16]. In addition, these shear bands are formed at frequencies in the range of 50–120 kHz when cutting speed was varying from 0.35 to 4.3 m/s and segment spacing becomes more periodic as cutting speed is increased. It is also noted that the dynamics of chip segmentation caused by thermal and elastic interaction between one and next shear zones can lead to aperiodic variations in the segment spacing [28].

Producing saw-tooth chips in orthogonal cutting of the 100Cr6 steel of HV730 hardness at cutting speeds of 25–285 m/min and feed rates of 0.0125–0.2 mm/rev was confirmed by Poulachon *et al.* [22]. Relevant mechanisms of formation of such chips, identified by means of a chip frozen technique (using a quick-stop device) are illustrated in Figure 3.26. According to Figure 3.26, four subsequent stages can be distinguished, namely:

1. During the tool indentation compressive stresses are induced in a zone around the cutting edge. Simultaneously, a crack at the workpiece surface at the angle Φ_f is initiated and it is followed by a shear plane which extends toward the cutting edge.
2. The chip volume of AA'BB' localized between the crack and the tool chamfer is ejected without deformation. The gap AA′ decreases progressively with the tool movement and the chip thickness h_c decreases. The temperature can increase up to the martensite transformation A_3 and a white layer may be produced.
3. The gap AA′ becomes so narrow that the material ejection velocity is very high and plastic deformation of the chip is very intensive. The second part of the chip of very small thickness is formed with extremely fast cooling.
4. The chip segment is formed and the gap AA′ fully disappears. The compressive stresses at the free surface are again high enough to produce a new crack. This periodic phenomenon will repeat itself leading to cyclic chip formation.

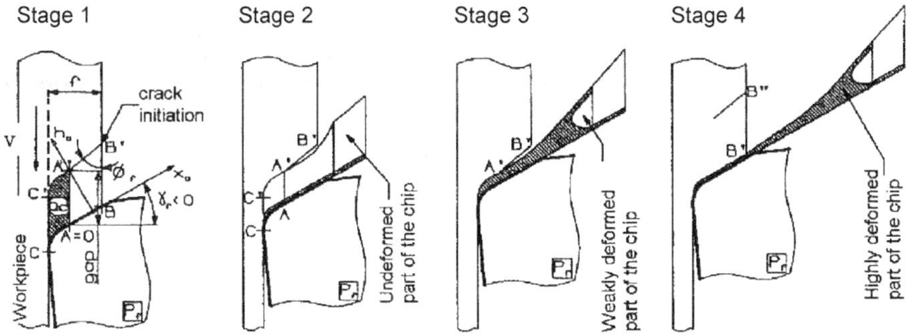

Figure 3.26 Stages of saw-tooth chip formation [21, 22]

3.2.4 Chip Morphology in Typical Machining Operations

It is well known that in machining processes chip morphology changes depending on the chip formation mechanisms (strain hardening *versus* thermal softening) but as discussed in Section 3.2.3 in hard machining saw-tooth type chips with different shapes and dimensions are prominently produced. In this section, the influence of cutting conditions and workpiece hardness on chip morphology will be addressed to hard-turning and milling operations. Special focus will be given to variable factors favouring saw-tooth chip formation.

Figure 3.27 depicts that depending on the workpiece hardness three characteristic values of the R criterion (Equation 3.2) can be determined and all possible chip morphologies are formed with the hardness ranging from 180 to 700 HV. In particular, the separating point of about 400 HV divides all chips produced into continuous shear-type chips ($R > 1$) and saw-tooth chips formed by cracking ($0 < R < 1$).

Figure 3.28 shows the longitudinal sections of chip produced for differently hardened AISI 4340 steel with CBN tools. It was validated that chips produced for the work material of lower hardness are continuous and the main formation

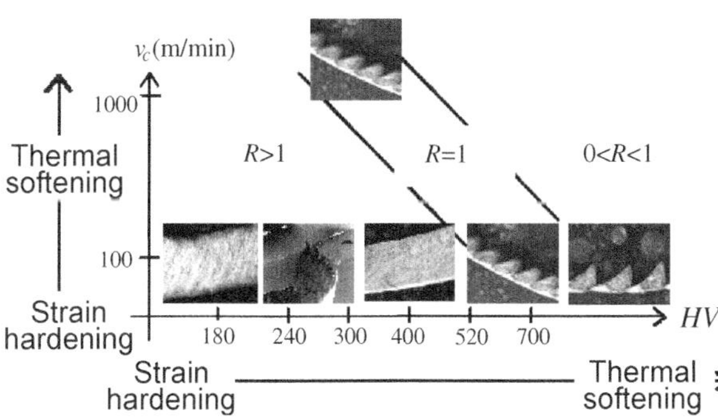

Figure 3.27 Variation of chip morphology with hardness [23]. Cutting parameters: v_c = 100 m/min, f = 0.1 mm/rev, a_p = 1 mm

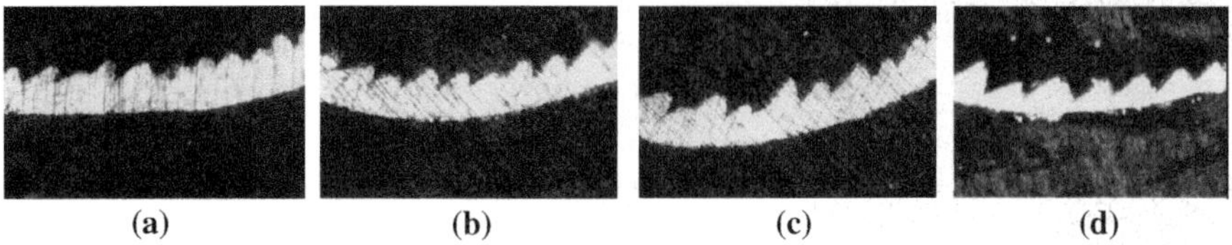

Figure 3.28 Types of the saw-tooth chips produced in turning of AISI 4340 steel of different hardness: (a) 45 HRC, (b) 50 HRC, (c) 55 HRC, and (d) 60 HRC [32]

mechanism is plastic deformation. However, when the hardness exceeds 45 HRC, the produced chips are thinner and their shapes change to saw-tooth type. As experimentally confirmed, the reduction of chip thickness results from the increase of the shear angle ($\Phi \cong 20°$ for 35 HRC and $\Phi \cong 40°$ for 60 HRC). Because heat is mostly transported into the chip, more heat is concentrated on the local shear bands, and a saw-tooth chips are formed.

Figure 3.29 shows the cross-sectional images of the chip top surface obtained in the milling of AISI H13 tool steel (50 ± HRC) with milling cutters equipped with (Ti, Al)N-TiN-coated carbide inserts. Two different types of chips are produced when keeping the following cutting parameters: $v_c = 100$–250 m/min, $f_z = 0.05$–0.20 mm/tooth, $a_e = 0.3$–0.6 mm, $a_p = 1.0$–2.5 mm; the near-uniform deformed continuous chips and the saw-tooth chips at higher cutting speeds and feeds. It can be concluded based on these observations that the chip morphology

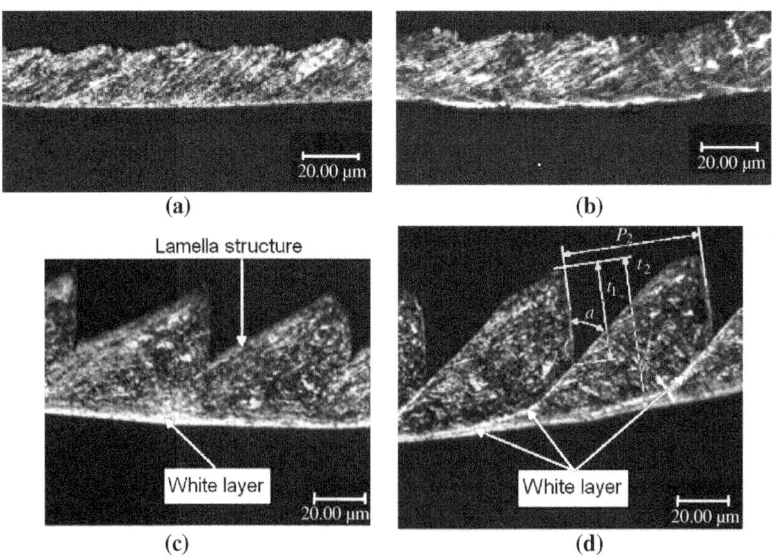

Figure 3.29 Combined effect of cutting speed and feed on cross-Section of chip top surfaces (p_c: segment pitch, t_1: saw height, t_2: chip thickness) [33]: (a) $v_c = 100$ m/min, $a_p = 2$ mm, $a_e = 0.6$ mm, $f_z = 0.05$ mm/tooth, (b) $v_c = 150$ m/min, $a_p = 2.5$ mm, $a_e = 0.4$ mm, $f_z = 0.10$ mm/tooth, (c) $v_c = 200$ m/min, $a_p = 2$ mm, $a_e = 0.4$ mm, $f_z = 0.15$ mm/tooth, and (d) $v_c = 250$ m/min, $a_p = 2.5$ mm, $a_e = 0.6$ mm, $f_z = 0.20$ mm/tooth

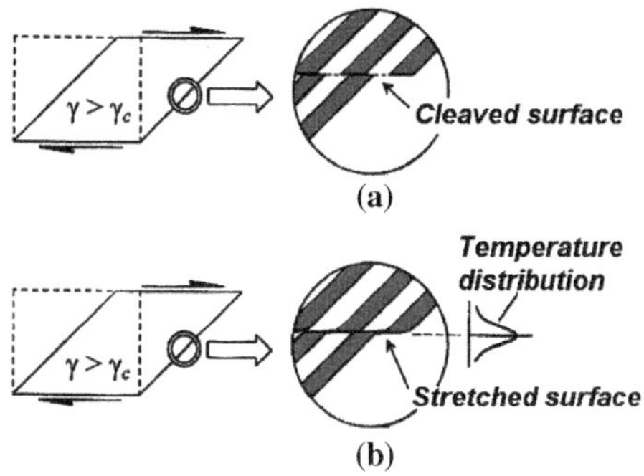

Figure 3.30 Illustration of the modes of material behaviour in hard machining: (a) lamella and (b) fold formation [27]

transition from continuous chip to segmented chip is favored by the combination of increased cutting speed and feeds. In particular, feed per tooth is the major contributor to chip serration, while cutting speed effect is the secondary one. Saw-tooth chips are also produced in turn-milling operations on hardened steels [34].

A key problem in hard machining is to define the transition from continuous to saw-tooth chips in terms of the cutting and work material parameters influencing the transition and the variation in chip morphology. According to Barry and Byrne [27] the primary instability occurring in the formation of saw-tooth chips is the initiation of adiabatic shear at the tool tip and propagation partway towards the free surface. Depending on the work material hardness and cutting conditions, catastrophic failure within the upper region of the primary shear zone occurs, through either ductile fracture or large-strain plastic deformation. It is postulated that the transition from the lamellar to fold structures of the free surface of chips (Figure 3.30, see also Figure 3.29 (c)) is part of the overall transition from continuous to saw-tooth chip formation, *i.e.* the initiation of adiabatic shear within the lower region of the primary shear zone. Figure 3.31 illustrates integrally the relationship between work material hardness, cutting conditions and chip morphology and the modes of plastic deformation within the primary shear zone.

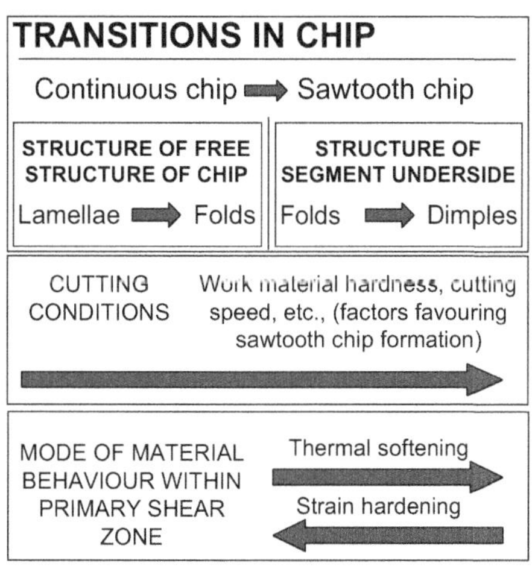

Figure 3.31 Transitions in chip morphology and the modes of plastic deformation in hard machining [27]

3.2.5 Material Side Flow Effect

The characteristic phenomenon occurring in hard-turning operations, termed a material side flow, is shown in Figure 3.32. This unique phenomenon of the plastic flow of brittle material is attributed to the squeezing effect of the workpiece material between the tool flank and the machined surface when the UCT is less than a minimum value h_{min}. Additionally, it can also result from flowing of highly plasticized material through the worn trailing edge to the side of the tool [35].

The material behaviour when the uncut chip thickness h is less than the minimum chip thickness h_{min} at point I, defined by the stagnation angle $\Theta \approx 25°$, is shown in Figure 3.32 (b). Because chip formation does not occur, elastic–plastic deformation of the surface layer is observed. At point II the elastic component Δh_{el} springs back after moving tool and behind point III the plastic deformation component Δh_{pl} leads to the final deformation of the surface layer.

Material side flow causes substantial deterioration of surface finish because the squeezed, flake-like, hard and very abrasive material is loosely attached to the generated surface along the feed marks. The formation of characteristic burrs on the feed mark ridges (Figure 3.33 (b)) is more intensive for higher cutting speed and larger tool nose radii, and when tool wear progresses.

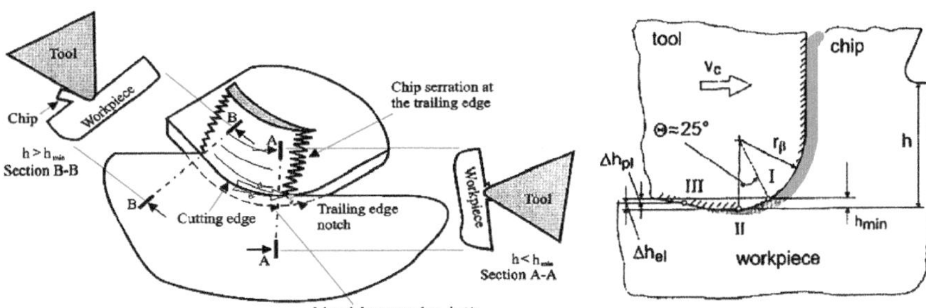

(a) (b)

Figure 3.32 Mechanisms of material side flow during hard turning [1, 36]

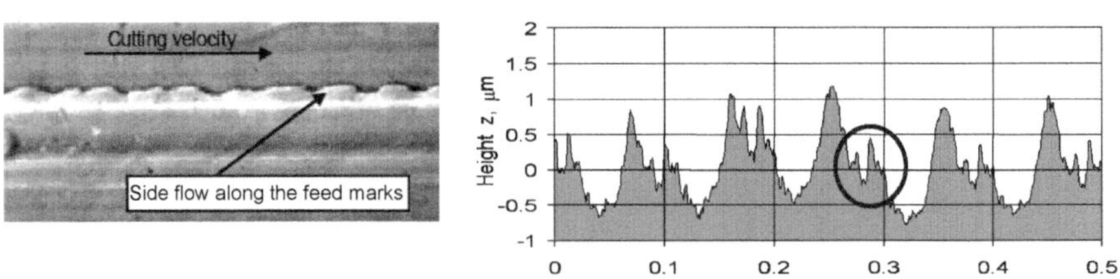

(a) (b)

Figure 3.33 (a) Scanning electron microscopy image showing burrs on a hard-turned profile due to material side flow [37], and (b) surface profile with characteristic later ridges (marked by circle) [35]

3.2.6 Finite-element-based Modelling of Chip Formation

This section overviews the state of numerical modelling using the FEM technique with special focus on hard-machining processes. The basic JC model and other modifications which are capable of simulating the two specific chip formation mechanisms, *i.e.* SSC and CTI (so far only CTI) can be considered in the FE model.

3.2.6.1 Constitutive Material Laws

In modelling of hard-machining processes using the FEM technique, the workpiece is usually modelled as a rigid–perfectly plastic body using the conventional JC material constitutive model in the form

$$\sigma = \left(A + B\varepsilon^{n}\right)\left(1 + C\ln\dot{\varepsilon}\right)\left[1 - \left(\frac{T - T_{0}}{T_{\text{melt}} - T_{0}}\right)^{m}\right] \tag{3.2}$$

where σ is flow stress, ε is effective stress, $\dot{\varepsilon}$ is effective strain rate and T is temperature in the deformation zone. Appropriate values of constants A, B, C and exponents n and m for hardened AISI 4340 and AISI 52100 can be found in [3] and [20].

Other FE models utilize the Von Mises criterion for computing the shear flow stress [21], BCJ (Baumann–Chiesa–Johnson) model which is capable of modelling adiabatic effects of polycrystalline materials [38] or Umbrello's hardness-based flow stress model [39, 40] given by Equation 3.3. A set of BCJ equations is given in [38].

$$\sigma\left(\varepsilon, \dot{\varepsilon}, T, \text{HRC}\right) = B(T)\left(C\varepsilon^{n} + F + G\varepsilon\right)\left[1 + \left(\ln\left(\dot{\varepsilon}\right)^{m} - A\right)\right] \tag{3.3}$$

where σ is flow stress, ε is effective stress, $\dot{\varepsilon}$ is effective strain rate and T is temperature in deformation zone, B is a temperature-dependent coefficient, C represents the work-hardening coefficient, J and K are two linear functions of hardness, A is a constant. Moreover, a non-isothermal viscoplastic-material model was applied. It is computed for AISI 52100 steel, $A = 0.0567$, $n = 0.083$, $C = 1092$ MPa, $m = 0.1259$, $F = 27.4$ HRC-1700.2 and $G = 4.48$ HRC-279.9 [20].

The model proposed by Huang and Liang [26] modified the JC equation by considering the dependence of the material flow stress on strain, strain rate and temperature according to Oxley's predictive theory, hence

$$\sigma = \left(\sigma_{0} + B\Phi^{n}\right) = \left(1 + C\ln\left(\frac{\dot{\varepsilon}}{\dot{\varepsilon}_{0}}\right)\right)\left(D - E\left(\frac{T - T_{\text{r}}}{T_{\text{m}} - T_{\text{r}}}\right)\right) \tag{3.4}$$

where C, B, D and E are constants and T_{r} is the reference temperature for measuring σ_0. This model is capable of predicting more severe deformation, *i.e.*

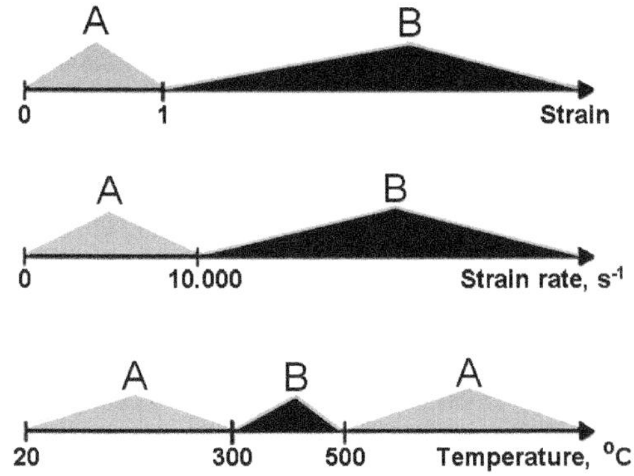

Figure 3.34 Schematic illustration of application ranges of different flow stress models in FEM simulation. *A*: model by Umbrello *et al.* [39]; *B*: model by Huang *et al.* [26]

when the effective strain and strain rate are high, but for intermediate temperatures. For example, it was incorporated in the two flow-stress model proposed by Kountanaya *et al.* [20] to represent the material behaviour in the primary shear zone, in which the temperature is relatively low.

Figure 3.34 illustrates application ranges of the different models, discussed previously, for predicting the flow shear stress in FEM simulation of hard turning.

3.2.6.2 Thermo-mechanical Instability

This problem was discussed in Section 3.2.2. In particular, Recht's criterion was defined and the cutting conditions for which shear instability occurs ($0 < R < 1$), were selected. The basic problem appearing in the modelling of chip formation in hard turning is to predict the critical cutting speed v_{cT} corresponding to the onset of shear instability, as suggested by Recht's criterion. According to Kountanaya *et al.* [20], a cutting speed of about 100 m/min is enough to dominate chip formation by CTI in hard turning of AISI 52100 steel (60 HRC).

3.2.6.3 Two- and Three-dimensional Chip Configurations

Figure 3.35 shows an example of a shear-zone model and a result of FEM simulation for which the saw-tooth chip is produced. Figure 3.36 shows the result of simulation of chip morphology based on the BCJ model in the case of turning of AISI 52100 steel (62 HRC) with a chamfered CBN tool under given cutting conditions. As can be seen in Figure 3.36, the saw-tooth chip, which is geometrically consistent with the measured chip, was predicted. It can be noticed that cyclic adiabatic shearing occurs inside the chip to form the serrated shear bands, which are distinctive characteristics of adiabatic deformation.

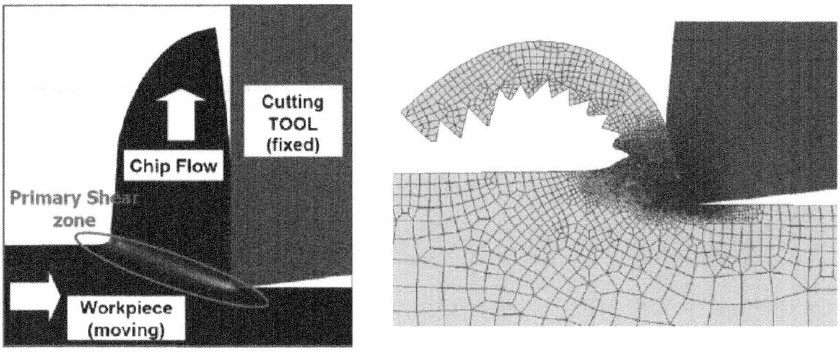

(a) (b)

Figure 3.35 Model of shear zone (a) and illustration of saw-tooth chip formation in hard turning (b) [20]

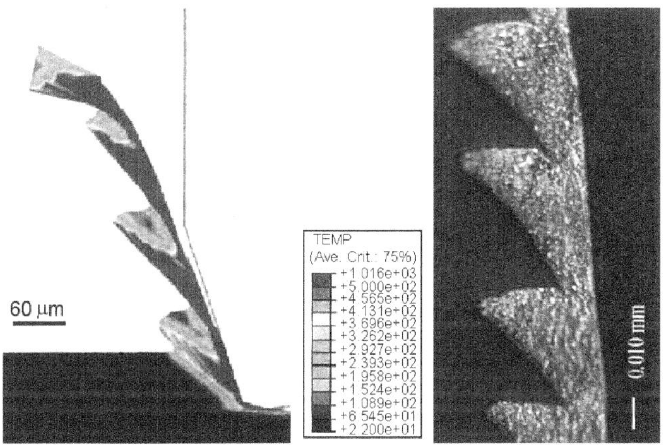

(a) (b)

Figure 3.36 Comparison of simulated (*a*) and real (*b*) chip morphologies for the BCJ model [38]. Cutting conditions: $v_c = 1.7$m/s, $a_p = 60$ μm, $\gamma_1 = 20°$, $\alpha_0 = 50$, $r_n = 8$ μm

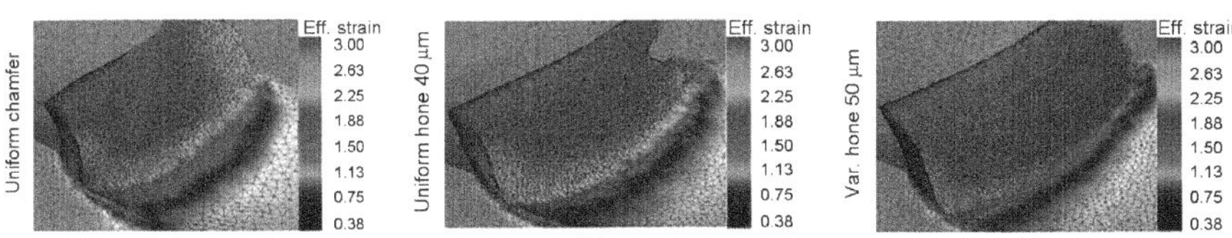

(a) (b) (c)

Figure 3.37 Simulated 3D chip formation for different micro-geometries of cutting tools: (a) uniform chamfer, (b) uniform hone, and (c) variable hone [4]

Figure 3.37 shows the effect of 3D simulation of chip formation in turning of AISI 4340 steel with PCBN tools of different cutting-edge geometries using the JC material model. As depicted, one segment of the saw-tooth chip is formed and uniformly chamfered cutting edges induced greater effective strains leading to higher thermo-mechanical loads on the workpiece materials.

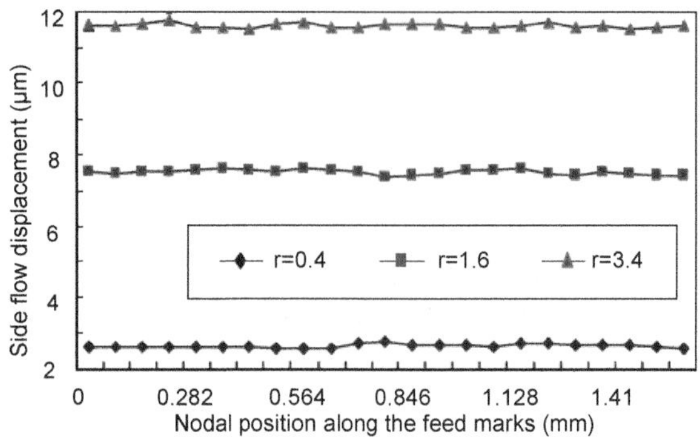

Figure 3.38 Influence of tool nose radius on material side flow along the surface profile [37]. Cutting conditions: $v_c = 1.18$ m/s, $f = 0.1$ mm/rev, $a_p = 0.25$ mm

3.2.6.4 Prediction of Material Side Flow

The material side flow generated during hard turning, shown in Figure 3.33, can also be modelled using a three thermo-elasto-viscoplastic material FE model. In [37] the Abaqus/Explicit™ package was used to predict chip formation and side flow effect in orthogonal turning of hardened AISI 52100 steel of hardness 62 HRC with a PCBN tool.

The model applied is capable of simulating the side flow of the material around the feed marks and, as shown in Figure 3.38, the side flow displacement depends on the value of tool nose radius in such a way that material is squeezed more laterally with larger tool nose radius (similarly for lower feeds).

References

[1] Grzesik W (2008) Advanced machining processes of metallic materials. Elsevier, Amsterdam
[2] Tönshoff HK, Arendt C, Amor R Ben (2000) Cutting of hardened steel. Ann CIRP 49(2):547–566
[3] Özel T, Karpat Y, Srivastawa A (2008) Hard turning with variable micro-geometry PcBN tools. CIRP Ann – Manuf Technol 57:73–76
[4] Özel T (2009) Computational modelling of 3D turning: Influence of edge micro-geometry on forces, stresses, friction and tool wear in PcBN tooling. J Mater Proc Technol 209:5167–5177
[5] Grzesik W, Wanat T (2006) Surface finish generated in hard turning of quenched alloy steel parts using conventional and wiper ceramic inserts. Int J Mach Tools Manuf 46:1988–1995
[6] www.coromant.sandvik.com; www.secotools.com
[7] Byrne G, Dornfeld D, Denkena B (2003) Advancing cutting technology. Ann CIRP 52(2):483–507
[8] Klocke F, Brinksmeier E, Weinert K (2005) Capability profile of hard cutting and grinding processes. Ann CIRP 54(2):557–580

[9] Ren H, Altintas Y (2000) Mechanics of machining with chamfer tools. Trans ASME J Manuf Sci Technol 122:650–659
[10] Klocke F, Kratz H (2005) Advanced tool edge geometry for high precision hard turning. Ann CIRP 54(1):47–50
[11] Ebrahimi A, Moshkar MM (2009) Evaluation of machinability in turning of microalloyed and quenched-tempered steels: Tool wear, statistical analysis, chip morphology. J Mater Proc Technol 209:910–921
[12] Elbestawi MA, Chen L, Becze CE, El-Wardany TI (1997) High-speed milling of dies and molds in their hardened state. Ann CIRP 46(1):57–62
[13] Gaitonde VN, Karnik SR, Figueira L, Davim JP (2009) Machinability investigations in hard turning of AISI D2 cold work tool steel with conventional and wiper ceramic inserts. Int J Refract Metals Hard Mater 27:754–763
[14] Grzesik W (2009) Wear development on wiper Al_2O_3-TiC mixed ceramic tools in hard machining of high strength steel. Wear 266:1021–1028
[15] Arsecularatne JA, Zhang LC, Montross C, Mathew P (2006) On machining of hardened AISI D2 steel with PCBN tools. J Mater Proc Technol 171:244–252
[16] Davies MA, Chou Y, Evans CJ (1996) On chip morphology, tool wear and cutting mechanics in finish hard turning. Ann CIRP 45(1):77–82
[17] Elbestawi MA, Srivastava AK, El-Wardany TI (1996) A model for chip formation during machining of hardened steel, Ann CIRP 45(1):71–76
[18] Varadarajan AS, Philip PK, Ramamoorthy B (2002) Investigations on hard turning with minimal cutting fluid application (HTMF) and its comparison with dry and wet turning. Int J Mach Tools Manuf 42:193–200
[19] Thepsonthi T, Hamdi M, Mitsui K (2009) Investigation into minimal-cutting –fluid application in high-speed milling of hardened steel using carbide mills. Int J Mach Tools Manuf 49:156–162
[20] Kountanya R, Al-Zkeri I, Altan T (2009) Effect of tool edge geometry and cutting conditions on experimental and simulated chip morphology in orthogonal hard turning of 100Cr6 steel. J Mater Proc Technol 209:5068–5076
[21] Ng EG, Aspinwall DK, Brazil D, Monagham J (1999) Modelling of temperature and forces when orthogonally machining hardened steel. Int J Mach Tools Manuf 39:885–903
[22] Poulachon G, Moisan A, Jawahir IS (2001) On modelling the influence of thermo-mechanical behavior in chip formation during hard turning of 100Cr6 bearing steel. Ann CIRP 50(1):47–50
[23] Guo YB, Liu CR (2002) 3D FEA modeling of hard turning. Trans ASME J Manuf Sci Technol 124:189–199
[24] Chen L, El-Wardany TI, Nasr M, Elbestawi MA (2006) Effects of edge preparation and feed when hard turning a hot work die steel with polycrystalline cubic boron nitride tools. CIRP Ann Manuf Technol 55(1):89–92
[25] Tönshoff HK, Arendt C, Ben Amor R (2000) Cutting of hardened steel. Ann CIRP 49(2):547–566
[26] Huang Y, Liang SY (2003) Cutting forces modeling considering the effect of tool thermal property – application to CBN hard turning. Int J Mach Tools Manuf 43:307–315
[27] Barry J, Byrne G (2002) The mechanisms of chip formation in machining hardened steels. Trans ASME J Manuf Sci Eng 124:528–535
[28] Davies MA, Burns TJ, Evans CJ (1997) On the dynamics of chip formation in machining hard metals. Ann CIRP 46(1):25–30
[29] Shaw MC, Vyas A (1998) The mechanism of chip formation with hard turning steel. Ann CIRP 47(1):77–82
[30] König W, Klocke F (1997) Fertigungsverfahren 1. Drehen, Fräsen, Bohren. Springer, Berlin
[31] König W, Berktold A, Koch KF (1993) Turning versus grinding – a comparison of surface integrity aspects and attainable accuracies. Ann CIRP 42(1):39–43
[32] Luo SY, Liao YS, Tsai YY (1999) Wear characteristics in turning high hardness alloy steel by ceramic and CBN tools. J Mater Proc Technol 88:114–121

[33] Zhang S, Guo YB (2009) An experimental and analytical analysis on chip morphology, phase transformation, oxidation, and their relationships in finish hard milling. Int J Mach Tools Manuf 49:805–813
[34] Schulz H, Kneisel T (1994) Turn-milling of hardened steel – an alternative to turning. Ann CIRP 43(1):93–96
[35] Grzesik W (2008) Machining of hard materials. In: Davim JP (ed) Machining. Fundamentals and recent advances. Springer, London, pp 97–124
[36] Kishawy HA, Elbestawi MA (1999) Effects of process parameters on material side flow during hard turning. Int J Mach Tools Manuf 39:1017–1030
[37] Kishawy HA, Haglund A, Balazinski M (2006) Modelling of material side flow in hard turning. Ann CIRP 55(1):85–88
[38] Guo YB, Wen Q, Woodbury KA (2006) Dynamic material behaviour modelling using internal state variable plasticity and its application in hard machining simulations. Trans ASME J Manuf Sci Technol 128:749–759
[39] Umbrello D, Rizzuti S, Outeiro JC, Shivpuri R, M'Saoubi R (2008) Hardness-based flow stress for numerical simulation of hard machining AISI H13 tool steel. J Mater Proc Technol 199:64–73
[40] Umbrello D, Fellice L (2009) Improving surface integrity in orthogonal machining of hardened AISI 52100 steel by modeling white and dark layers formation. CIRP Ann Manuf Technol 58:73–76

Chapter 4
Surface Integrity

A.M. Abrão, J.L.S. Ribeiro and J.P. Davim

Surface integrity comprises the study of the alterations induced during the manufacture of a component that might affect its properties and service performance. Therefore, additionally to geometric irregularities (surface texture and both dimensional and geometric deviations), the study on surface alterations (such as metallurgical alterations, cracks and residual stresses) induced by hard-part machining is of utmost importance, especially in the case of components subjected to dynamic loading. Consequently, this chapter is focused on the investigation of the influence of tool material and geometry and cutting parameters on the surface integrity of components subjected to hard-part machining and, when applicable, comparisons are drawn with grinding and non-conventional processes, especially electrical discharge machining (EDM).

4.1 Geometric Irregularities

Owing to the fact that in most cases hard-part machining is employed as an alternative to grinding, the requirements concerned with surface texture (microgeomet-

---

A.M. Abrão
Department of Mechanical Engineering, Universidade Federal de Minas Gerais,
Av. Antônio Carlos, 6627, Pampulha, 31270901, Belo Horizonte MG, Brazil
e-mail: abrao@ufmg.br

J.L.S. Ribeiro
Department of Mechatronic Engineering, Pontifical Catholic University of Minas Gerais,
Av. Dom José Gaspar, 500, Coração Eucarístico, 30535-610, Belo Horizonte MG, Brazil
e-mail: ribeirojls@yahoo.com.br

J.P. Davim
Department of Mechanical Engineering, University of Aveiro, Campus Santiago,
3810-193 Aveiro, Portugal
e-mail: pdavim@ua.pt

ric irregularities) and dimensional and geometric deviations (macrogeometric irregularities) are rather tight. The irregularities generated after hard machining are largely affected by the cutting parameters and tool geometry as well as by the machine-tool condition. According to [1], high-precision machining of steels requires two basic developments: improvement of the levels of accuracy obtainable using available hard-turning technology and the enlargement of ultraprecision technology for the machining of ferrous materials, especially the use of monocrystalline cutting materials. Conventional computer numerical control precision lathes equipped with roller bearing spindles and conventional bed ways are not suitable for high-precision machining of hardened steels. In addition to high rigidity, the machine-tool design must take into account that the passive force is the principal component, in contrast to the machining of alloys of lower hardness, where the tangential (cutting) force is the principal component of the machining force.

4.1.1 Surface Finish

Feed rate (f) and tool nose radius (r_ε) are widely recognized as the principal parameters affecting the surface finish of turned components. Rech and Moisan [2] assessed the influence of cutting speed (v_c) and feed rate on the surface finish of 27MnCr5 steel case hardened to 850 $HV_{0.3}$ at a depth of 0.6 mm when turning with polycrystalline cubic boron nitride (PCBN) compacts. In spite of the low surface roughness obtained at a feed rate of 0.05 mm/rev (R_a values below 0.2 μm), plastic flow caused by high temperature and pressure was observed on the freshly machined surface. Therefore, the authors recommend a minimal feed rate between 0.05 and 0.1 mm/rev in order to avoid side flow. Finally, the R_{max} parameter was found to be more sensitive to the evolution of tool wear than R_a.

The influence of tool material (PCBN with high and low CBN content, mixed alumina, whisker-reinforced alumina and silicon-nitride based) and cutting speed on the surface roughness when finish turning AISI H13 hot-work die steel (average hardness of 52 HRC) and AISI E52100 bearing steel (62 HRC) was investigated by [3]. The low-CBN-content and the mixed-alumina tools provided superior surface finish for both work materials (R_a values as low as 0.14 μm), although in the case of the tool steel this was achieved at the highest cutting speed ($v_c = 200$ m/min) and for the bearing steel at the lowest speed ($v_c = 70$ m/min).

Javidi *et al.* [4] studied the influence of the above-mentioned parameters (feed rate ranging from 0.05 to 0.4 mm/rev and nose radii of 0.2–0.4 and 0.8 mm) on the maximum surface roughness (R_{max}) obtained after turning a hardened carbon steel. The results indicated that the difference between the actual and theoretical values increased with the elevation of the latter.

Jacobson *et al.* [5] reported that when turning a hardened bainite steel with PCBN tools at cutting speeds varying from 50 to 999 m/min best surface finish ($R_a = 1.7$ μm) was recorded using a cutting speed of $v_c = 170$ m/min. A comparison between the performance of cutting fluids when turning hardened AISI 4340 steel

(49 HRC) with mixed-alumina tools was carried out by [6]. The findings indicated that when finishing at higher cutting speeds ($v_c = 300$ and 400 m/min), the use of cutting fluids resulted in lower scatter in the roughness of the machined surface. In general, the emulsion promoted better surface finish compared with the synthetic cutting fluid, which was occasionally outperformed by dry machining.

The influence of tool material and cutting time when finish turning ($v_c =$ 180 m/min, $f = 0.08$ mm/rev and $a_p = 0.15$ mm) case-hardened DIN 19MnCr5 steel (surface hardness of 66 HRC) with mixed-alumina, TiN-coated mixed-alumina and PCBN tools is shown in Figure 4.1. In addition to slightly different chamfer preparations, the alumina tools presented a nose radius of $r_\varepsilon = 0.8$ mm and the PCBN, $r_\varepsilon = 0.4$ mm, therefore, lower R_a values are expected to be observed when the alumina tools are used. Comparing both alumina inserts it can be noticed that the coated ceramic provided better surface finish, probably due to its superior wear resistance, which maintained the integrity of the cutting edge for a longer period. In spite of the higher R_a values promoted by the lower tool nose radius, the PCBN tool experienced lowest wear rate. Consequently, this grade promoted the least increase in surface roughness as cutting time elapsed (28 % against 46 % for the mixed alumina and 44 % for the coated alumina).

Benga and Abrão [7] employed the surface response methodology in order to identify the optimal machining parameters (cutting speed and feed rate) responsible for lower surface roughness values when turning DIN 100Cr6 bearing steel quenched and tempered to an average hardness of 62 HRC using mixed alumina, whisker-reinforced alumina and two grades of PCBN tools. Comparable surface finish values were achieved using the four tool grades (lowest roughness of $R_a = 0.25$ μm), nevertheless, while the lowest feed rate resulted in lowest surface roughness, in the case of cutting speed the optimum for the mixed-alumina and PCBN inserts was obtained for cutting speed ranging from $v_c = 116$ to 130 m/min. Using cutting speeds below this range, the temperature rise was not sufficient to reduce the shear strength of the work material and, consequently, the cutting

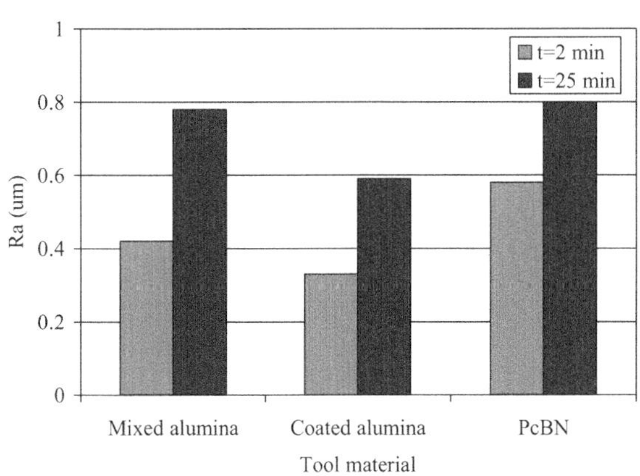

Figure 4.1 Effect of tool material and cutting time on surface roughness of case-hardened DIN 19MnCr5 after finish turning ($v_c = 180$ m/min, $f = 0.08$ mm/rev, and $a_p = 0.15$ mm)

forces, whereas cutting speeds above this range lead to machine-tool vibration, which can impair surface finish. In the particular case of the whisker-reinforced alumina, lowest surface roughness was produced at the lowest cutting speed tested ($v_c = 100$ m/min), probably due to the high tool wear rates observed when cutting speed was elevated, which led to the deterioration of the cutting edge.

The influence of workpiece hardness on the surface finish obtained after turning AISI 4340 steel quenched and tempered to 42 and 50 HRC with, respectively, coated carbide and PCBN inserts was studied by [8]. In spite of the higher tool wear rate observed for the PCBN tool, this grade promoted best surface finish owing to its larger nose radius compared with the coated carbide insert ($r_\varepsilon = 1.6$ mm against $r_\varepsilon = 0.8$ mm) and to the higher cutting forces recorded when turning the softer material (caused by the larger seizure area present when machining the steel of lower hardness). In addition to that, surface roughness values in the range of $R_a = 0.28$–1.12 μm were obtained when turning AISI D2 steel (58 HRC) at cutting speeds from 80 to 220 m/min, feed rates from 0.05 to 0.15 mm/rev and constant depth of cut of 0.2 mm, best surface finish recorded using the highest cutting speed and lowest feed rate.

The performance of mixed-alumina tools with conventional and wiper geometry used for turning hardened AISI D2 cold-work die steel (60 HRC) was compared by [9–10]. The ceramic tool with wiper geometry promoted better surface finish than the conventional geometry, in spite of the higher turning forces observed when the former geometry was tested.

High-speed endmilling trials were performed by [11] in order to investigate the influence of cutting speed and tool material (indexable carbide, cermet and PCBN inserts and solid carbide) on the surface finish of AISI D2 cold-work tool steel (58 HRC). The findings indicated that surface roughness (R_a parameter) did not follow any particular trend when cutting speed was elevated and that the influence of tool wear on the roughness of the machined surface was higher when the cutter with indexable carbide inserts was employed. Finally, surface roughness values in the range of $R_a = 1$–6 μm were obtained using a carbide ball-nose endmill, while the PCBN endmill produced considerably better surface finish ($R_a = 0.1$–0.2 μm).

Elbestawi *et al.* [12] investigated the surface finish obtained after high-speed milling AISI H13 hot-work die steel hardened to 45 and 55 HRC. PCBN inserts with high and low CBN content were brazed on solid carbide ball-nose endmills with different edge preparations. Cutting speeds of 220 and 1320 m/min were employed (rotational speeds of 10,000 and 60,000 rpm, respectively). Using the lower cutting speed, surface roughness varied from $R_a = 0.2$ to 0.6 μm as tool wear progressed and lower R_{max} values were obtained for the harder material (2 μm for 55 HRC and 4 μm for 45 HRC). However, increasing cutting speed resulted in poorer surface finish ($R_a = 0.65$ μm) owing to the vibration of the magnetic spindle and an opposite trend with regard to the effect of workpiece hardness on surface roughness (higher R_{max} values recorded for the harder material).

Figure 4.2 shows the influence of cutting speed and feed rate per tooth (f_z) on the surface finish obtained after endmilling quenched and tempered AISI H13 steel (average hardness of 47 HRC) using coated carbide inserts. It can be seen in

Figure 4.2 (a) that at a cutting speed of $v_c = 120$ m/min surface roughness reaches its lowest value, probably for the reasons previously discussed by [7]. As feed rate per tooth increases, see Figure 4.2 (b), surface roughness is elevated drastically owing to the higher distance between peaks and valleys. A more discrete effect is observed when depth of cut (a_p) is altered (Figure 4.2 (c)), *i.e.*, higher R_a values are recorded when a_p is increased due to the larger shear plane area, which requires higher milling forces. Finally, Figure 4.2 (d) presents the influence of cutting fluid applied under distinct machining conditions. When milder cutting parameters are selected, the roughness values are similar, however, under more severe cutting conditions it seems that the use of cutting fluid presents a detrimental effect on surface finish. One possible reason for this behaviour may be the increase in the thermal shock when cutting fluid is applied, which may accelerate fatigue wear of the tool and impair surface finish.

As far as drilling of hardened steels is concerned, the amount of information available in the published literature is considerably lower compared with turning and milling. A number of reasons can be listed as constraints to drilling of hard materials: variable cutting speed tending to zero at the centre of the drill, which results in extruding (instead of shearing) the work material by the chisel edge; variable rake, clearance and inclination angles together with small cross-section

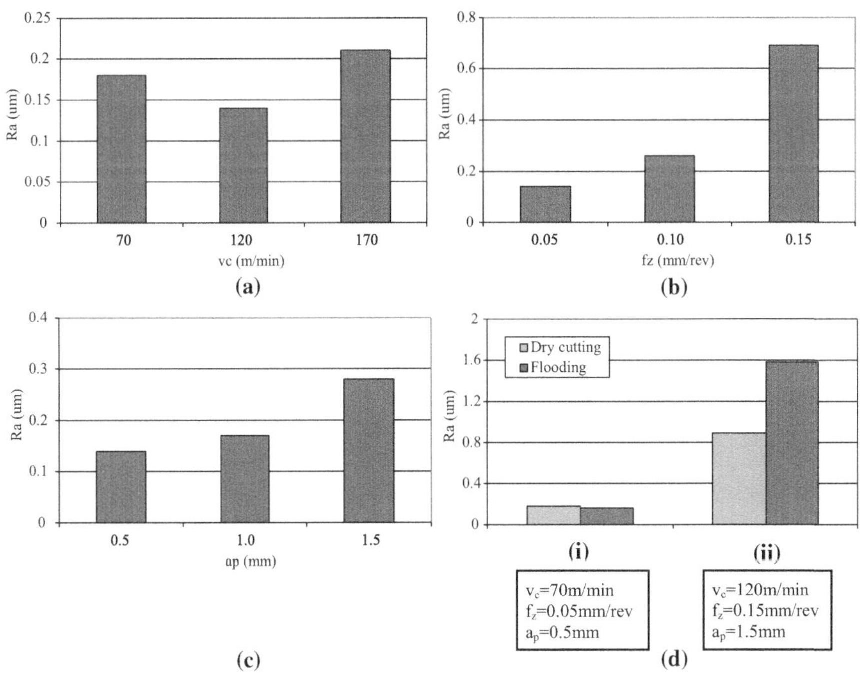

Figure 4.2 Effect of cutting conditions on surface roughness of AISI H13 tool steel (47 HRC) after endmilling using coated carbide inserts: (a) cutting speed; (b) feed rate; (c) depth of cut; (d) cutting fluid ((i) $v_c = 70$ m/min, $f_z = 0.05$ mm/rev, $a_p = 0.5$ mm; (ii) $v_c = 120$m/min, $f_z = 0.15$ mm/rev, $a_p = 1.5$ mm)

area of the drill, leading to insufficient torsional rigidity to cut this grade of materials. Despite the above-mentioned drawbacks, surface roughness values within $R_a = 0.14–0.48$ μm were obtained by [13] after drilling AISI H13 steel (hardness range of 54–55 HRC) with coated solid carbide drills (8 mm diameter) at cutting speeds from 20 to 45 m/min and feed rates of 0.1 and 0.2 mm/rev.

The development of computer-controlled three-dimensional profilemeters in recent decades has allowed the use of this technique to characterize machined surfaces more realistically. Waikar and Guo [14] compared the topography of turned and ground hardened AISI 52100 steel (62 HRC) and reported that, additionally to the regular pattern produced on the surface by the combination of feed rate and tool nose radius when turning (against the random pattern observed after grinding), in general the surface produced by gentle grinding presented lower-amplitude parameters (S_a, S_q, S_p, S_v, S_t, and S_z) compared with turning using fresh and worn inserts and with dry grinding. As far as the functional parameters skewness (S_{sk}) and kurtosis (S_{ku}) are concerned, only the surface generated by gentle grinding presented negative S_{sk}, thus indicating superior fluid retention ability, while only abusive grinding produced a surface with S_{ku} higher than three (height distribution with low standard deviation).

The influence of the cutter path orientation on the surface texture of AISI H13 hot-work die steel (52 HRC) subjected to high-speed milling with coated tungsten carbide ball-nose endmills was investigated by [15]. The angle between the workpiece and the horizontal was 75° and the single raster and raster cutter strategies were tested using the following orientations: vertical upward and downward and horizontal upward and downward cutting. The findings indicated that the single raster strategy with vertical upward cutter orientation gave best surface texture. When a new cutter was employed, this condition produced well-defined isotropic surfaces, whereas lowest surface roughness was obtained when a worn cutter was used. Tool vibration, high milling forces and side flow were typical events that impaired the quality of the surfaces produced by the other strategies.

The effect of cutting speed (40 and 60 m/min), tool wear (flank wear $VB_B = 0$ and 0.2 mm) and cutting fluid pressure (ranging from 5 to 450 bar) on the surface roughness obtained after turning Inconel 718 with carbide tools was investigated by [16], who reported that R_a values within 0.7 to 1 μm were obtained using a new tool, irrespective of the remaining parameters. Not surprisingly, this value increased with tool wear to reach values as high as $R_a = 3$ μm.

Turning trials on Inconel 718 (44 HRC) using ceramic and PCBN inserts with various geometries (square, round and triangular) and two edge preparations (sharp edge with chamfer of 20° × 0.1 mm and honed edge with chamfer of 15° × 0.15 mm) were conducted by [17]. Lowest surface roughness values were obtained using the round insert, due to the fact that its radius was 6 mm, while the nose radius of the square and triangular inserts was $r_\varepsilon = 0.8$ mm. In general, the mixed-alumina insert provided lower R_a values compared with the SiC-reinforced alumina and PCBN tools, as well as the use of the honed edge.

Similar work was carried out by [18], who investigated the influence of the geometry of coated carbide tools on the surface finish of solution annealed and precipitation hardened Inconel 718 (35 HRC) after face-turning trials at v_c = 60 m/min, f = 0.1 mm/rev and a_p = 0.5 mm. In general, lower surface roughness was provided by the round inserts compared with the square inserts for the reason previously explained (larger nose radius), as well as for inserts with honed edge in comparison with sharp and chamfered edges. The reason for the poorer surface finish obtained using chamfered tools was attributed to the higher cutting forces and, in the case of sharp tools, excessive chipping of the cutting edge. With regard to the rake angle, positive values provided better surface finish when wet cutting, however, a negative rake angle allowed lower R_a values when dry turning. Finally, the use of cutting fluid resulted in superior surface finish owing to the absence of a built-up edge.

Sharman *et al.* [19] assessed the effect of tool geometry and coating on the surface texture of holes produced on Inconel 718 and found scattered R_a values within a range from 1 to 1.5 µm regardless of alterations on geometry and coating. However, as tool wear progressed, significant smearing was observed on the hole walls. Subsequent reaming and boring improved surface finish drastically, providing average R_a values below 0.5 µm.

4.1.2 Dimensional and Geometric Deviations

The static and dynamic stiffness of the machine-tool/workpiece pair are of great importance, since vibration must be minimized during cutting. This can be achieved through the proper design of bed ways and both headstock and tailstock bearings. Additionally, high accuracy of the critical elements of the machine tool is required in order to produce components with quality comparable with grinding.

The influence of tool wear (measured in terms of length cut) on the dimensional and geometric deviations after turning case-hardened DIN 19MnCr5 steel using PCBN (v_c = 180 m/min, f = 0.08 mm/rot and a_p = 0.15 mm) are presented in Figures 4.3 and 4.4, respectively. Figure 4.3 shows that the dimensional deviation for a given dimension observed during the first 1600 mm is minimal (corresponding to ISO tolerance grade IT5); however, it increases drastically after this point, probably due to the wear of the cutting tool.

As far as the geometric deviations are concerned, see Figure 4.4, distinct trends are observed for roundness and concentricity, *i.e.*, while a minimal variation of 4 µm is noted for roundness, the cylindricity deviation increases continuously with length cut. Apparently, tool wear did not affect roundness to the same extent that it affected concentricity, while the machine-tool condition could play a more relevant role in concentricity.

Matsumoto *et al.* [20] compared the roundness deviation of bearing elements (58–62 HRC) subjected to hard turning with PCBN and grinding (both followed by superfinishing) and found that, owing to the accuracy of the machine tools employed, tighter tolerances were obtained after hard turning, although the deviations generated by operations were within the component tolerance range.

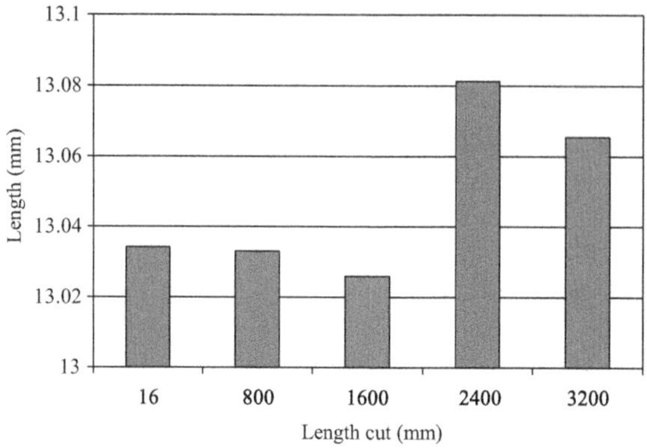

Figure 4.3 Effect of length cut on dimensional deviation when hard turning DIN 19MnCr5 steel using PCBN ($v_c = 180$ m/min, $f = 0.08$ mm/rot, and $a_p = 0.15$ mm)

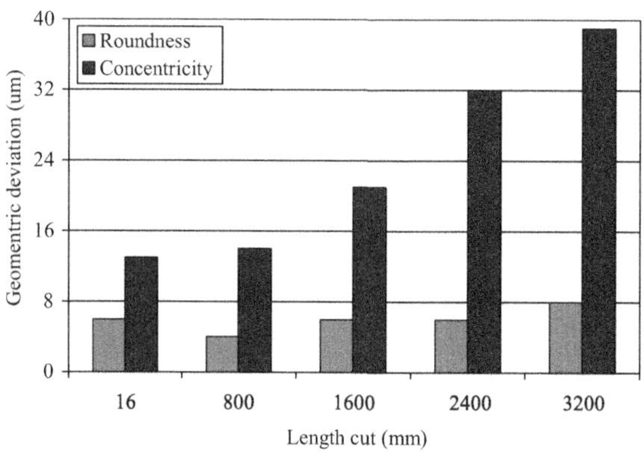

Figure 4.4 Effect of length cut on roundness and concentricity when hard turning DIN 19MnCr5 steel using PCBN ($v_c = 180$ m/min, $f = 0.08$ mm/rot, and $a_p = 0.15$ mm)

Figures 4.5 and 4.6 show, respectively, the dimensional and geometric deviations observed after milling ($v_c = 120$ m/min, $f = 0.1$ mm/rot, $a_p = 0.5$ mm and $a_e = 5$ mm) circular pockets on hardened H13 tool steel (41 HRC) using different strategies and indexable tools. It can be seen in Figure 4.5 that the actual diameter of the cavities was smaller than its nominal value (70 mm) due to tool deflection and that rather wide tolerances were produced.

Figure 4.5 (a) shows that lower dimensional deviation is obtained when the up-milling strategy is employed. The reason for that may reside in the higher plastic deformation caused by the impact at the teeth entry when down-milling. As far as the tool material is concerned, see Figure 4.5 (b), slightly lower deviation was produced by the cermet, probably due to its superior wear resistance.

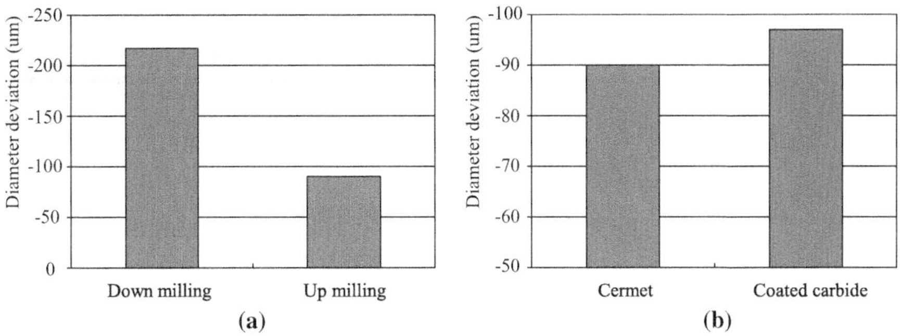

Figure 4.5 Dimensional deviations obtained after milling circular pockets on hardened H13 steel: (a) effect of strategy, and (b) effect of tool material

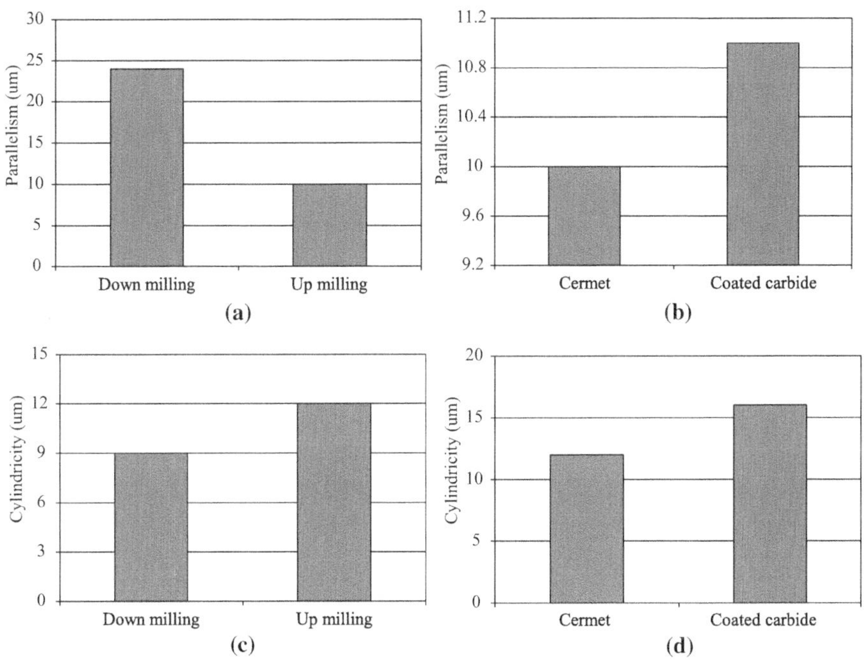

Figure 4.6 Geometric deviations obtained after milling circular pockets on hardened H13 steel: (a) effect of strategy on parallelism; (b) effect of tool material on parallelism; (c) effect of strategy on cylindricity; (d) effect of tool material on cylindricity

The same behaviour noted for the dimensional deviation was observed for parallelism (Figure 4.6 (a) and (b)), *i.e.*, tighter tolerances were obtained when up-milling with cermet; however, when considering the cylindricity deviation, down-milling promoted a slightly better result than up-milling (Figure 4.6 (c)) and again the cermet outperformed the coated carbide tool (Figure 4.6 (d)).

A geometric error of 25 µm was observed by [12] after high-speed milling hardened hot-work die steel using PCBN inserts brazed on solid carbide ball-nose endmills. As tool wear progressed, the error increased to 75 µm.

Coldwell *et al.* [13] investigated the quality of holes produced on hardened AISI H13 hot-work steel (54–55 HRC) after drilling with AlTiN-coated solid carbide tools with 8 mm diameter. Using an emulsion as cutting fluid (5 % concentration), a diameter deviation of ±5 µm was obtained. Cylindricity values increased with cutting speed, the best value being obtained for $v_c = 20$ m/min (20.4 µm). With regard to the roundness deviation, the values ranged from 2.85 to 17.75 µm.

4.2 Surface Alterations

According to Bellows and Tishler, cited in [21], there are five types of surface alterations related to metal removal: mechanical (which include plastic deformation, built-up edge defects, hardness alteration, cracks, residual stresses, voids, pits, burrs and inclusions), metallurgical (phase transformation, grain size and distribution, precipitate size and distribution, foreign inclusions, twinning and recrystallization, among others), chemical (such as intergranular attack, corrosion and oxidation, preferential dissolution, contamination, enbrittlement by chemical absorption and pits), thermal (including heat-affected zone and recast layer) and electrical (conductivity and magnetic changes and resistive heating and overheating). In spite of the fact that this classification is quite useful and detailed, some alterations cannot be classified in just one type. For instance, hardness alteration may occur as a result of plastic deformation (mechanical type) and/or phase transformation (metallurgical alteration).

Nevertheless, surface and subsurface alterations induced by hard-part machining are typically of mechanical, metallurgical and thermal nature. The presence and extent of the alterations depend greatly on the severity of the machining operation. The principal alterations observed on the surface layers of components subjected to hard-part machining are presented below.

4.2.1 *Microstructural Alterations*

Microstructural changes in steels may be either mechanically or thermally induced. In the case of hardenable steels, if the workpiece temperature exceeds the austenization temperature during machining (due to friction and plastic strain), austenite will be formed and, after quenched by the cold bulk material or by the cutting fluid, a brittle, highly stressed and crack-prone martensite layer (usually called a white layer) is formed at the surface. However, if the workpiece temperature exceeds the tempering temperature only, then overtempering will take place, leading to the softening of the affected layers (identified as a dark layer after etching). Rech and Moisan [2] and Bosheh and Mativenga [22] claim that the formation of the white layer is a thermal process involving phase transformation and, probably plastic deformation, which has not been fully understood yet.

Turning tests on hardened AISI H13 tool steel (54–56 HRC) with PCBN inserts indicated that the hardness and depth of the white layer decreased as cutting speed was elevated owing to a slight reduction in the workpiece temperature [22]. In contrast, Axinte and Dewes [23] assert that neither a white layer nor a heat-affected zone were observed after high-speed milling AISI H13 hot-work die steel hardened to 47–49 HRC with TiAlN coated carbide ball-nose endmills. This difference can be explained by the fact that in the former work cutting speeds of 100, 400 and 700 m/min were tested, the thickest white layer being observed at the lowest cutting speed. For cutting speeds of 400 and 700 m/min, the depth of the white layer decreased dramatically. In the case of the latter work, cutting speeds of 200 and 300 m/min were used, which seem to be above the critical value required for the formation of the white layer.

Figure 4.7 presents cross-section samples of AISI H13 tool steel (47 HRC) after dry endmilling under gentle conditions (Figure 4.7 (a)), grinding with a CBN wheel (Figure 4.7 (b)) and die-sinking EDM using copper electrodes (Figure 4.7 (c)). Similarly to the work reported by [23], the samples subjected to milling and grinding do not show any evidence of alteration. Conversely, cracks, recast layer and heat-affected zone are visible in the sample subjected to EDM. As a consequence, one would expect a similar behaviour of the components produced through endmilling and grinding with regard to their service life, both outperforming the part produced by EDM.

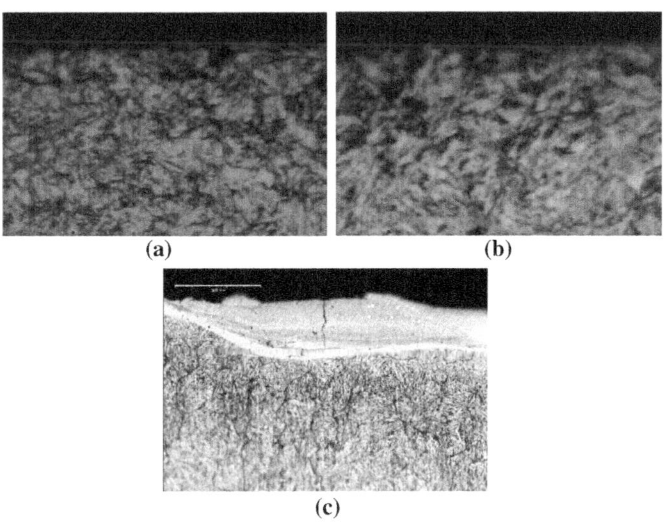

Figure 4.7 Microstructure of hardened AISI H13 after machining using distinct operations: (a) endmilling (×1000), (b) grinding (×1000), and (c) die sinking EDM (×500)

The occurrence of the white layer in AISI 52100 bearing steel hardened to 63 HRC after turning with PCBN and grinding with an alumina wheel under abusive cutting conditions was studied by [24]. In order to promote the generation of the white layer, the turning and grinding tests were conducted dry using a worn PCBN tool and a worn alumina wheel, respectively. Both operations resulted in a white layer followed by a dark layer before the bulk material, however, the thickness of the layers was considerably larger when the abrasive process was employed. Additionally, the white layer produced by turning was etch resistant, while the microconstituents were visible in the sample generated by grinding (ferrite matrix plus cementite particles without evidence of severe strain). The authors claim that the etching resistance of the turned sample may be attributed to the presence of nanograins (100–300 nm) resulting from dynamic recovery and recrystallization.

Furthermore, the volume fraction of retained austenite in the white and dark layer was 10.64 % for the turned sample and 2.88 % for the ground sample, whereas in the dark layer the values found were 11.68 % and 0 %, respectively. The high volume of retained austenite observed in the white layer of the turned specimen suggests that severe plastic strain took place during machining. Moreover, these findings indicate that the temperature reached in the ground dark layer was high enough for tempering, but not to form austenite.

The volume fraction of retained austenite on the surface of hard-turned cold-work steel is reported to be twice the amount found in the bulk material [25]. The authors state that the generation of a white layer after machining is the result of a combination of the thermal effect and plastic deformation. Under circumstances in which the austenisation temperature is not reached, the thermal effect causes the softening of the material, thus leading to severe plastic deformation and the formation of a nanocrystalline white layer. As a result of the softening, an overtempered martensitic layer is observed below the white layer.

The influence of the cutting-edge preparation on the plastic deformation induced by the cutting tool when turning hardened AISI 52100 bearing steel (average values of 41 and 57 HRC) was investigated by [26]. Low-content PCBN inserts with three distinct edge preparations were tested: sharp (average honed edge radius of $r_h = 22.9\,\mu m$), honed edge ($r_h = 100–150\,\mu m$) and chamfered ($115\,\mu m \times 17°$ plus $r_h = 25.4\,\mu m$). The metallographic analysis indicated that subsurface plastic flow occurred only after machining using the insert with larger hone edge ($r_h = 100–150\,\mu m$). Furthermore, deeper subsurface plastic deformation was associated with the presence of larger compressive residual stresses.

A direct relationship between the presence of the white layer on the machined subsurface and both edge preparation and depth of cut was reported by [27], who noticed that when turning AISI 52100 bearing hardened to 60–62 HRC with PCBN inserts, the combination of large edge hone radius ($r_h = 70\,\mu m$) and high depth of cut ($a_p = 0.255$ mm) results in the formation of an untempered martensite layer followed by an overtempered martensite layer, not present either after turning with tools possessing low edge radius ($r_h = 25\,\mu m$) at low depth of cut ($a_p = 0.051$ mm), or after grinding and superfinishing.

After turning hardened AISI 52100 bearing steel (62 HRC), Abrão and Aspinwall [28] found a white untempered martensite layer 2 µm deep, followed by an overtempered martensite layer; however, cracks, tears, laps and recast layer were not evident. Matsumoto *et al.* [29] did not observe any significant structural changes on the cross-sections of AISI 4340 steel (54 HRC) samples subjected to milling, with plastic deformation taking place very near the surface. According to these authors, under normal circumstances, *i.e.*, when there is not excessive tool wear or when abusive cutting conditions are not employed, martensite formation is not expected to happen.

Javidi *et al.* [4] noticed that the extent of the plastic deformation obtained after hard turning 34CrNiMo6 steel was approximately 3–4 µm, irrespective of alterations in feed rate and tool nose radius. According to [12], the depth of the damaged layer induced after high-speed milling AISI H13 with PCBN brazed endmills varied from 4 to 6 µm and depends on tool wear and edge preparation (sharp, honed with 0.025 mm radius and chamfered with 20°). The influence of feed rate and depth of cut was considered negligible.

Sharman *et al.* [16] report that plastic deformation in the cutting direction, carbide cracking and surface cavities are alterations typically induced after turning Inconel 718; however, more severe plastic deformation is observed when worn tools are used in comparison with fresh inserts. In contrast, differences in the level of microstructural deformation were not evident when cutting fluid was applied using conventional flooding and high pressure (from 70 to 450 bar). Drilling the same material, a white layer was present in addition to plastic deformation (Sharman *et al.* [19]).

4.2.2 Hardness Alterations

In general, hardness measurement are carried out with Vickers or Knoop diamond indenters using the average indented diagonal in the former or the longer diagonal in the latter. Due to the fact that hardness may vary considerably within a short distance, loads below 1 N (microhardness test) are often employed. However, in contrast to the high loads typically employed for hardness testing, microhardness value depends on the applied load. Tönshoff and Brinksmeier [30] compared Knoop and Vickers indenters and noticed that the former presents the following advantages: the longer diagonal reduces the risk of misreading, the influence of lack of homogeneity normal to the surface is reduced due to the smaller indentation depth, the small diagonal allows measurement near the machined surface and it can be used to measure anisotropy. On the other hand, the Vickers indenter is less affected by form errors on the surface and the indenter is cheaper. Additionally, the above-mentioned authors state that owing to the steep variation in microhardness values, especially near the machined surface, measuring should be performed using the slope method, in which the indentations are produced on an tapered surface, thus allowing the measurement of hardness close to the machined surface (not possible when the sample is cross-sectioned due to the fact that a distance of not less than 2.5 times the diagonal indentation must be kept from a disturbed area). Finally, the polishing method employed prior to microhardness testing may affect the results: higher hardness values and wider scatter are obtained when grinding the samples in comparison with electrolytic polishing, probably due to microstructure non-homogeneity and work hardening induced by the former procedure.

A relationship between nanohardness and both phase transformation and residual-stress distribution was established by [31] after turning AISI 52100 bearing steel hardened to 62 HRC with fresh and worn PCBN tools. The results indicated that higher nanohardness values were obtained when the worn insert was employed, probably due to the presence of a thin white layer (5–10 μm deep). Moreover, when compressive residual stresses are present, the slope of the nanoindentation loading curve at initial yielding tends to increase, whereas tensile or less compressive stresses tend to decrease the slope of the loading curve.

García Navas *et al.* [25] assessed the nanohardness variation recorded after hard turning AISI O1 tool steel with worn inserts and noticed that high values are obtained on the surface, decreasing steeply to values below the bulk hardness just below the surface and returning to the bulk value at a depth of 20 μm. In contrast, when the same work material is wire-electrical-discharge machined, lower hardness values are recorded on the surface, decreasing smoothly to reach the bulk hardness 20 μm below the surface.

The influence of the operation and tooling on microhardness variation of AISI 52100 bearing steel (62 HRC) after dry turning with PCBN and mixed-alumina (Al_2O_3 + TiC) inserts was studied by [28]. The results indicated that after turning, a maximum hardness value of approximately 900 $HV_{0.025}$ was obtained near the surface, decreasing to a minimum 4 μm below, followed by an increase to the bulk hardness (750 $HV_{0.025}$) at approximately 10 μm beneath the surface. In addition to that, the lowest hardness value found using the PCBN tool was inferior to that obtained turning with the mixed-alumina tool.

Microhardness (Knoop indenter at a load of 25 gf) and nanohardness measurements (Berkovich indenter at a maximum load of 8 mN) were carried out on hardened bearing steel samples subjected to turning and grinding followed by superfinishing [32]. Both methods indicated that the ground specimens presented higher hardness values than the turned samples, probably due to the size effect induced by the small down-feed employed in the grinding operation, which leads to a severe stress gradient near the machined surface.

Hashimoto *et al.* [32] report that the apparent softening measured at the surface using the microhardness method is not observed when the nanohardness technique is employed. This behaviour can be observed in Figure 4.8, where it can be noticed that lower microhardness values were recorded near the surface of endmilled hardened AISI H13 hot-work die tool (47 HRC) using different machining conditions, although the hardness values remained practically unaltered beneath the surface (average microhardness of 492.7 $HV_{0.1} \pm 3.7$ in Figure 4.8 (a) and 493.3 $HV_{0.1} \pm 5.4$ in Figure 4.8 (b)). Considering that the microstructural

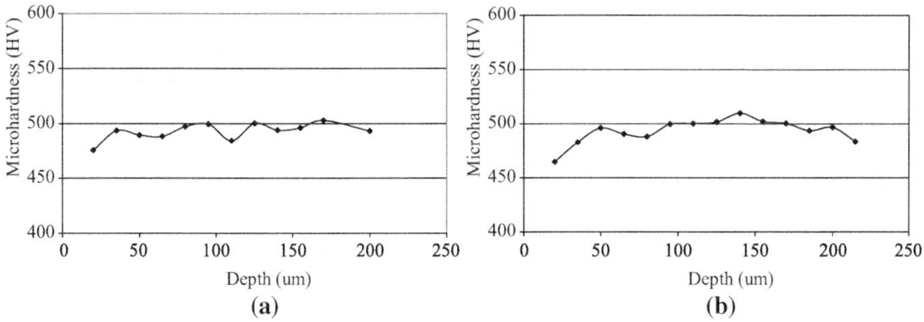

Figure 4.8 Effect of the cutting parameters on microhardness variation of AISI H13 steel after endmilling: (a) $v_c = 70$ m/min, $f_z = 0.05$ mm, $a_p = 0.5$ mm and $a_e = 20$ mm (dry), and (b) $v_c = 120$ m/min, $a_p = 1.5$ mm, $f_z = 0.15$ mm and $a_e = 20$ mm (flooding)

analysis of the respective cross-sections is similar to that presented in Figure 4.7 (a), *i.e.*, does not show evidence of a heat-affected zone, and that the microhardness testing was conducted using a load of 100 gf, it is likely that the lower microhardness values observed near the surface are due to the edge effect.

Guo and Sahni [24] compared the microhardness profile of hardened bearing steel after abusive turning and grinding and noticed that the microhardness of the white layer generated on the ground samples was approximately 40 % higher than those observed on the turned specimens and that the dark layer induced by grinding was thicker.

The influence of tool wear and cutting-fluid pressure on the microhardness of Inconel 718 was investigated by [16]. The findings indicated that when turning with a new tool the microhardness profile did not change drastically (400–420 $HK_{0.05}$), whereas when a worn tool was employed, the highest microhardness value was recorded at the surface (480 $HK_{0.05}$) and decreased to 420 $HK_{0.05}$ at a depth of 100 μm. However, using cutting fluid at high pressure (450 bar) did not promote any significant alteration on the microhardness profile. According to the authors, the reason for this behaviour may be related to the reduction in both tool wear and friction between the worn tool and the workpiece surface when highly pressurized cutting fluid is applied.

Microhardness values measured near the surface of high-speed-turned Inconel 718 were found to be approximately 50 % higher than those of the bulk material [33]. The microhardness profiles indicated that a considerable amount of plastic deformation was induced by machining at a maximum depth of 200 μm below the surface.

4.2.3 Residual-stress Distribution

Residual stresses induced by machining operations can be assessed directly using, for instance, the X-ray diffraction method to measure the distance between planes of atoms, or indirectly, employing strain gauges or optical, electronic or mechanical displacement transducers to determine the deformation induced when the stresses are relieved.

The residual stresses induced on a component are the result of a combination of mechanical and thermal effects. In general, the mechanical action (burnishing) leads to plastic deformation and promotes compressive residual stresses. The thermal effect (temperature rise due to friction and plastic strain) however, may promote tensile or compressive residual stresses depending on the maximum temperature reached at the workpiece and corresponding microstructural changes that take place. For instance, the transformation from austenite to martensite during rapid cooling (by the bulk material, cutting fluid or air) involves a volume expansion caused by the change from a face-centred cubic structure to a more open tetragonal structure, thus resulting in compressive stresses in the surface layers. The layers beneath the surface, however, reach lower temperatures and cool at slower rates, therefore, their contraction is restrained by the higher strength of the layers above. Consequently, tensile residual stresses may be induced below the machined surface. The resulting residual stress depends on the magnitude of the mechanical and thermal effects, nevertheless phase transformation induced by cutting can be neglected as a cause of residual stresses on hardened steels [34].

Koster, according to [35], states that when surface roughness lies in the range $R_a = 2.5–5$ μm, residual stress is often a better indicator of fatigue performance than surface topography. Additionally, this effect is reduced as temperature is elevated owing to the relaxation of residual stress with the exposure of the workpiece to heat.

The blind-hole drilling method was employed by [4] in order to measure the residual stresses induced after turning a quenched and tempered carbon steel. The results indicated that only compressive residual stresses were obtained. Additionally, the magnitude of the compressive stresses increased with feed rate and decreased as tool nose radius was elevated.

A number of authors have tried to establish relationships between the magnitude and depth of residual stresses and the machining parameters, aiming at optimization of the latter. Choi [36] used the Taguchi method to investigate the influence of the machining parameters (cutting speed, feed rate, depth of cut and cutting fluid) and tool geometry (nose radius and chamfer angle) on the residual stresses (measured using the X-ray diffraction method) and rolling contact fatigue life of hardened AISI 1053 steel after facing with PCBN inserts. Compressive residual stresses were induced with peaks at 5–20 μm from the surface. Moreover, the maximum residual stress value ranged from –400 to –1600 MPa.

Gunnberg *et al.* [37] employed a face-centred-cubic experimental design to assess the influence of the machining parameters on the residual stresses induced by hard turning of DIN 18MnCr5 case-carburized steel (hardness of 550 HV at a depth of 1.2 mm) using low-content PCBN tooling. An increase in cutting speed promoted tensile residual stresses on the surface only, owing to the fact that the heat generated by higher cutting speeds was unable to reach the layers beneath the surface. The elevation of feed rate induced compressive residual stresses and the influence of depth of cut was found to be negligible. Finally, the more negative the rake angle of the tool, the more compressive the residual stresses on the surface and at depths up to 50 μm below the surface.

Similar work was carried out by [38], who investigated the influence of the tool rake angle and feed rate on the residual stresses induced on AISI 52100 bearing steel hardened to 62 HRC after turning with PCBN tooling. Under all circumstances, tensile stresses were observed on the machined surface and shifted to compressive beneath the surface. As the rake angle decreased from –6 to –61°, the magnitude and depth of the compressive stresses were elevated. In addition to that, the intensity of the compressive residual stress increased as feed rate was elevated and was not affected by depth of cut.

Xueping *et al.* [39] reported that compressive residual stresses are obtained after hard face turning a bearing steel (62–63 HRC) with PCBN tooling. The authors employed the Taguchi method in order to find the optimal cutting parameters that would promote compressive stresses of higher magnitude. The findings indicated that the cutting parameters affected the residual stresses in the circumferential and radial directions in different manners: when considering the circumferential direction, cutting speed was the most significant factor, followed by feed rate and depth of cut. When the radial direction was analysed, cutting speed remained the principal parameter followed, however, by depth of cut and feed rate. Additionally, the lowest level for average stresses was obtained when the specimens were machined at the lowest cutting speed ($v_c = 30$ m/min) and highest depth of cut ($a_p = 0.135$ mm), irrespective of the stress-measuring direction. As far as the feed rate is concerned, lowest average stress in the circumferential direction was obtained using the highest feed rate value ($f = 0.25$ mm/rev), whereas lowest stress in the radial direction was found using the intermediate feed rate value ($f = 0.15$ mm/rev), although this effect was regarded as minimal by the authors.

Conversely, Matsumoto *et al.* [20] did not notice any significant influence of depth of cut on the circumferential residual stress induced after turning with PCBN followed by superfinishing case-carburized bearing steel (58–62). Furthermore, increasing feed rate resulted in tensile residual stresses of higher magnitude on the surface without a clear trend in the layers beneath. Considering that the materials tested by [39] and [20] presented equivalent hardness values and that in both cases PCBN tooling was employed and the stresses were measured using the X-ray diffraction technique, the difference in the results can be attributed to the superfinishing operation employed in the latter and/or to the heat treatment used: quenching and tempering in the former work or carburizing in the latter.

According to [2], the higher the cutting speed, the higher the module of the tensile residual stress. The influence of feed rate was found to be negligible compared with cutting speed and the use of TiN-coated PCBN promoted higher compressive stresses in comparison with an uncoated tool, probably owing to the superior tribological behaviour of the coating, which reduces friction and, as a consequence, cutting temperature.

The influence of cutting speed (from 50 to 999 m/min) on the residual-stress distribution after hard turning a bainitic steel (58 HRC) with PCBN tools at a constant feed rate of $f = 0.1$ mm/rev was investigated by [5], who noticed that compressive residual stresses were induced; nevertheless, the maximum compressive stress was recorded when machining at a cutting speed of $v_c = 230$ m/min. According to the authors this phenomenon can be explained by the fact that the strain rate increases with cutting speed, promoting mechanical work and resulting in compressive residual stresses. A further increase in cutting speed results in more heat, which in turn increases temperature and produces residual stresses.

Compressive residual stresses were induced after turning AISI 52100 bearing steel [28] with fresh and worn PCBN and mixed-alumina inserts. The difference between the tool grades was not significant; nevertheless, the intensity and depth of the stresses was higher when worn inserts (average flank wear $VB_B = 0.27$ mm) were tested (maximum residual stress of approximately –600 MPa at a depth of 20 µm below the surface). This can be explained by the fact that using worn tools requires higher cutting forces to shear the work material, thus inducing compressive stresses, whereas lower mechanical energy is necessary when new cutting tools are used.

García Navas *et al.* [25] reported that tensile residual stresses were induced on the surface of hardened AISI O1 cold-work tool steel after turning with a worn tool. The residual stresses tended to be compressive below the surface and became zero at approximately 300 µm deep. Similarly, tensile residual stresses were obtained on the surface of the work material after wire EDM, nevertheless they tended to zero at 80 µm below the surface.

A comparison between the profile of the circumferential residual stresses obtained after hard turning and grinding case carburized bearings was carried out by Matsumoto *et al.* [20]. Both operations produced compressive residual stresses on the bearings surface, however, in the case of the turned bearings the magnitude of the compressive stress increased up to 40 µm beneath the surface before decreasing to zero, whereas for the ground samples the stress magnitude decreased drastically, tending to zero at 20 µm below the surface.

Similar results were reported by [27], who observed that both turning with PCBN tools and grinding with Al_2O_3 wheels induced compressive residual stresses on hardened AISI 52100 steel (60–62 HRC). However, while the peak stress took place at the surface for the ground specimens and decreased steeply, in the case of the turned samples the maximum value was recorded at a depth of 10–20 µm and decreased smoothly. Moreover, subjecting turned and ground samples to further surperfinishing promoted an increase in the magnitude of the residual stresses within a layer 20 µm thick.

When turning AISI 4340 steel (hardness range from 29 to 56 HRC) using mixed-alumina tools, Matsumoto *et al.* [34] noticed that the residual stresses shifted from tensile to compressive as the workpiece hardness increased. In addition to that, the maximum compressive stress and corresponding depth of the affected layer increased with workpiece hardness.

Similar results were observed by [2], who reported that the residual stresses on the machined surface shifted from compressive to tensile as tool wear progressed when turning a case-hardened steel (850 $HV_{0.3}$) with PCBN inserts. The magnitude of the compressive stresses recorded beneath the surface was reduced with tool wear, probably due the higher cutting temperatures measured as tool wear increases, thus leading to the formation of martensite and, consequently, to tensile stresses.

Matsumoto *et al.* [29] noticed that the profile of the residual stress induced after milling AISI 4340 steel hardened to 54 HRC with mixed-alumina inserts did not change considerably when measured parallel or perpendicular to the cutting direction. Nevertheless, the magnitude of the compressive stress was higher when cutting parallel to the specimen length in comparison to a perpendicular cut, reaching –600 MPa at a depth of 10 μm below the machined surface. Additionally, samples with similar R_a values but distinct R_{max} values behaved in a different manner during the fatigue tests (longer fatigue life observed for the specimens with lower R_{max} values).

The effect of the machining parameters on the residual stresses promoted by high-speed milling AISI H13 tool steel (47–49 HRC) with coated carbide tools was investigated by [23]. Highly compressive stresses were obtained and the significant factors were cutting speed, feed rate and workpiece angle. Increasing cutting speed and feed rate caused the compressive stresses to decrease due to an increase in the thermal effect. In the case of an elevation in the workpiece angle, the residual stress decreased due to the absence of the mechanical effect (rubbing of the centre of the ball-nose mill against the machined surface).

Matsumoto *et al.* [20] studied the influence of the cutting-edge geometry on the residual stresses induced by hard turning a bearing steel and found that using PCBN inserts with a 0.2-mm honed edge, higher and deeper compressive residual stresses were induced in comparison with a chamfered insert, probably due to the higher amount of plastic deformation promoted by honing. A similar effect was produced when a double-chamfered insert was compared with a single-chamfered tool.

Analogous work was carried out by [26], who studied the effect of the edge preparation on the residual-stress distribution (measured through the X-ray diffraction technique) obtained after hard turning AISI 52100 bearing steel with PCBN tools. The performance of three distinct edge preparations was compared: sharp (average honed edge radius $r_h = 22.9$ μm), honed edge ($r_h = 100–150$ μm) and chamfered (115 μm × 17° plus $r_h = 25.4$ μm) and the results indicated that the honed edge induced the deepest and most intense residual stresses, probably owing to the fact that this geometry increased the friction interaction between tool and work material.

Liu *et al.* [40] studied the influence of tool nose radius ($r_\varepsilon = 0.4–0.8$ and 1.2 mm) on the residual-stress distribution when finish turning JIS SUJ2 bearing steel (60 HRC) with PCBN inserts. The elevation in tool nose radius caused the residual stress on the machined surface to shift from compressive to tensile stress. Moreover, the magnitude of the peak compressive stress observed beneath the surface was reduced as tool nose radius was increased. Furthermore, as tool wear progresses, the compressive stresses recorded on the machined surface shift to tensile, while the magnitude of the compressive stresses observed beneath the surface increase, *i.e.*, the stress amplitude increases with tool wear. In addition to that, the depth affected increases from approximately 50 to 200 μm using a worn tool.

The effect of tool wear on the residual stresses distribution observed when turning case-hardened steel (62 HRC) with PCBN inserts was reported by [1]. The results indicated that residual stresses are not observed on the surface machined with a fresh tool and compressive residual stresses were observed 20 μm below the surface. However, as tool wear progresses, tensile stresses of increased magnitude are observed on the machined surface, while higher compressive stresses are recorded below the surface. These findings suggest that friction due to tool wear is the principal cause of tensile residual stresses.

Figure 4.9 outlines the influence of the principal processes used for metal removal on hard materials on the residual-stress distribution. In spite of the fact that the intensity and depth affected by the residual stress may vary according to the measuring direction, machining parameters employed, tool material and edge preparation, tool wear, composition and hardness of the work material, this graph summarizes the outcome of the thermal and mechanical effects, which will induce, respectively, tensile or compressive stresses. It can be noticed that the peak residual stress (either of tensile or compressive nature) usually takes place at some distance from the machined surface, except for high-speed machining (HSM), which induced maximum compressive residual stress on the surface.

In general, abusive grinding and EDM are responsible for tensile stresses only. These operations require a large amount of energy to provide low metal removal rate, therefore, high machining temperatures are generated. In addition to that, the low mass of the chip prevents heat being conducted away from the workpiece, thus promoting a further temperature increase. Conversely, the swarf plays a vital role in conveying heat from the cutting zone when turning.

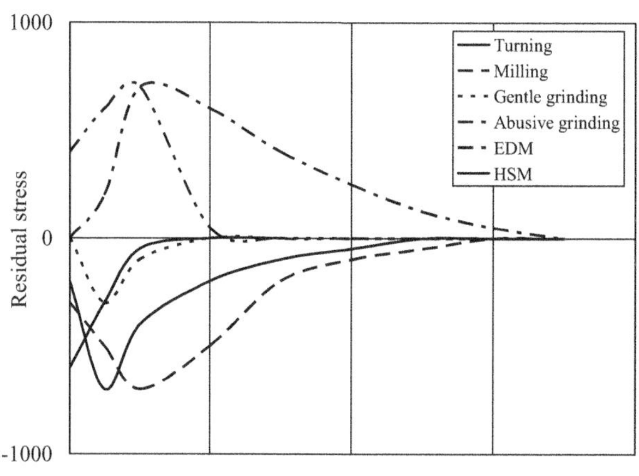

Figure 4.9 Influence of the machining operation on the residual-stress distribution in hardened steels

A comprehensive investigation on the influence of the cutting parameters (cutting speed, feed rate and depth of cut) and edge preparation (chamfered 30° × 0.1 mm with and without edge honing and chamfered 20° × 0.1 mm) on the residual-stress distribution of high-speed turned Inconel 718 was conducted by [33]. PCBN tools were employed and the Taguchi method was used to identify the optimal parameters. In spite of the fact that the effect of none of the parameters was found to be statistically significant within a confidence level of 95 %, the results suggest that residual stresses decrease, *i.e.*, become more compressive, at high cutting speed (v_c = 400 m/min), low feed rate (f = 0.05 mm/rev), high depth of cut (a_p = 1 mm) and using chamfered tools (30° × 0.1 mm) with honed edge. The reason for such behaviour resides in the ability of the chip in dissipating heat, especially in materials with poor thermal conductivity, such as Inconel 718. As cutting speed is elevated, the rate of heat dissipation by the chip increases and, consequently, tensile residual stresses shift to compressive stresses. The relationship between heat generation and dissipation is again used to explain the influence of feed rate and depth of cut on the residual-stress distribution. As far as the edge preparation is concerned, the use of tools with geometry that promotes larger contact area with the workpiece (chamfered and honed) is expected to induce compressive residual stresses.

Coelho *et al.* [17] compared the residual stress on the machined surface of turned Inconel 718 (44 HRC) and found compressive residual stresses when round mixed-alumina tools were used. Additionally, the intensity of the compressive stress increased when a honed edge was employed, in comparison with a sharp edge. In contrast, tensile residual stresses were observed on the surface of Inconel 718 after turning with fresh carbide tools, thus decreasing steeply to the lowest compressive-stress value approximately 50 μm below the surface and returning to the bulk value at a depth of 100 μm [16]. When worn tools were employed under conventional flooding, the magnitude of the tensile stress increased to a maximum value of 1650 MPa, as well as the depth required to reach the bulk value (200 μm). However, when the cutting fluid was applied at high pressure (450 bar) on the rake face of the tool, compressive stresses were not recorded. Finally, a similar behaviour was observed when flood cooling (5 bar) and applying cutting fluid at high pressure on the flank face.

The influence of tool geometry and cutting fluid on the surface residual stresses induced after face turning Inconel 718 (35 HRC) with coated carbide inserts was investigated by Arunachalam *et al.* [18]. According to the authors, compressive residual stresses are induced when round inserts with chamfered edge and negative rake angle are used, owing to the higher amount of plastic deformation under these circumstances. The use of cutting fluid resulted in either compressive or slightly tensile residual stresses. When dry cutting with sharp tools possessing a positive rake angle, however, tensile residual stresses were recorded due to the dominance of the thermal effect. The use of round inserts and wet machining presented the same beneficial effect on residual stresses when PCBN tools were employed [41]. Additionally, mixed-alumina tools induced tensile stresses, probably due to the low thermal conductivity of the ceramics associated to the demand for dry machining when this grade is used.

Differently from hardened steels, the residual-stress distribution in the hoop and radial directions promoted by face turning a nickel-based alloy with carbide insert differ drastically [42]. While the residual stress in the hoop direction is tensile on the surface and decreases monotonically, the stress measured in the radial direction is tensile on the surface too, albeit of lower magnitude, and changes to compressive beneath the surface. Furthermore, the authors investigated the influence of tool geometry and coating and noticed that higher tensile stresses were obtained using a round insert in comparison to a rhombic tool, while the presence of coating did not affect residual stress significantly. As far as tool damage is concerned, an increase in tool wear promoted higher tensile stresses on the machined surface, while chipping of the cutting edge resulted in compressive stress with high magnitude (–1000 MPa) and penetration depth of 400 μm.

Post-machining treatments such as shot peening, burnishing and abrasive tumbling tend to induce compressive residual stresses and, consequently, to improve the fatigue strength. The stress depth is in the range from 100 to 300 μm for shot peening, whereas glass bead peening promotes a compressive layer with depth ranging from 25 to 75 μm depending on the process parameters [36].

In the particular case of shot peening, Segawa *et al.* [43] claim that despite the fact that this operation can change tensile residual stresses to compressive stresses, it presents the following drawbacks: it is an additional and time-consuming process and it is difficult to maintain the accuracy of complex components. These authors propose a burnishing tool to be used simultaneously with a cutter during the milling operation in order to induce compressive residual stresses on the machined part. The cutter is similar to a three-edge endmill (6 mm diameter) and possesses a central burnishing pin made of tungsten carbide (3 mm diameter) that protrudes 0.1 mm from the cutting edges. The findings indicated that compressive residual stress values comparable to those obtained after shot peening were recorded.

4.2.4 Fatigue Strength

The fatigue process consists of three stages: initial fatigue damage leading to crack nucleation and crack initiation in regions where the strain is most severe, progressive growth of the crack (crack propagation) and finally sudden fracture of the remaining cross-section [44]. Czyryca [45] states that many testing devices and specimen designs are available for fatigue testing according to the mode of loading: direct (axial) stress, plane bending, rotating beam, alternating torsion or combined stress. The selected loading mode should replicate, as accurately as possible, the actual service condition of the sample being tested.

Comparing the fatigue life of SUJ JIS2 bearing-steel specimens (62 HRC) subjected to turning with PCBN and mixed-alumina tools, Abrão and Aspinwall [28] noticed that longer fatigue life was obtained for the samples turned using the

PCBN compact, although a run-out was obtained for both cutting tools when a stress range of approximately 900 MPa was applied. Similar results were obtained by [29], who found an endurance limit of 890 MPa for AISI 4340 steel (54 HRC) subjected to milling with mixed-alumina tools.

The influence of tool nose radius on the fatigue behaviour of hardened 34CrNiMo6 steel was investigated by [4]. Surprisingly, the finding indicated that turning the samples with the lowest tool nose radius and, consequently, generating the highest surface roughness, resulted in highest fatigue strength. Such behaviour is explained by the fact that highest compressive residual stresses are induced when the work material is machined with tools possessing the lowest tool nose radius, thus suggesting that the residual stress presents a more pronounced influence on the fatigue behaviour than surface finish.

According to [36], who conducted rolling contact fatigue tests on specimens produced using various machining conditions, the residual stresses induced on the machined component possess a significant effect on its contact fatigue life. Additionally, tool nose radius was the principal parameter affecting the performance of the samples, *i.e.*, longest fatigue life was obtained when the specimens were machined using cutting tools with the smallest nose radius ($r_\varepsilon = 0.79$ mm).

Rolling contact fatigue tests comparing the performance of two batches of case-carburized bearing-steel (58–62 HRC) specimens subjected to hard turning with PCBN and grinding (both followed by superfinishing) were performed by [20]. The findings indicated that the samples subjected to hard turning presented a performance similar or superior to the ground specimens. However, in the case of AISI 52100 bearing steel quenched and tempered to 61–62 HRC, an increase of twofold in the rolling contact fatigue life was observed for the specimens subjected to turning followed by superfinishing [32].

Guo *et al.* [46] compared the fatigue lives of hardened bearing-steel (62 HRC) specimens subjected to turning with PCBN and grinding with alumina wheels. Surface compressive stresses were induced in both sets of specimens, which were free of white layer. The results indicated that the fatigue life of samples obtained through turning was approximately 40 % longer compared with that obtained after grinding. However, when testing turned and ground specimens with both surface tensile stresses and white layer, the rolling contact fatigue life decreased drastically and there was no difference between the fatigue lives of turned and ground samples.

The fatigue behaviour of hardened AISI 52100 steel (60–62 HRC) specimens subjected to turning with PCBN and grinding with Al_2O_3 was assessed by [27]. Tension–tension axial tests were performed and the results indicated that the fatigue life of the hard-turned specimens outperformed those produced by grinding. The presence of a white layer on the hard-turned specimens did not affect their performance in the fatigue tests. In addition to that, superfinishing of turned and ground specimens improved drastically the performance of the former (increase in fatigue life of 470 %), while its effect on the latter was marginal (35 %).

4.3 Conclusion

Knowledge on the influence of the machining operation, cutting parameters and tool condition on the surface integrity of hard metallic alloys is of utmost importance for the performance of the finished component. In addition to the geometric irregularities (surface texture and tolerances), changes on the surface and subsurface layers of the machined part must be assessed and controlled. The following aspects are highlighted:

- Surface finish comparable to grinding is reasonably achievable when machining hardened steels. In addition to the factors that traditionally affect surface roughness, *i.e.*, feed rate and tool nose radius, special attention should be paid to the deterioration of the cutting edge when hard machining, owing to the fact that accelerated wear rates are observed, especially when the tool grade is not properly selected, thus impairing the surface finish of the component. In the particular case of milling, the use of solid carbide cutters results in better surface roughness in comparison with indexable inserts. When cutting nickel alloys, inferior surface roughness values are obtained compared with hardened steels. Edge preparation seems to significantly affect surface texture of nickel alloys, best results being obtained using a honed edge.
- As far as the dimensional and geometric deviations are concerned, the machine-tool design and stiffness are critical to achieve tight tolerances. Tolerance range close to that provided by grinding is occasionally obtained when turning difficult-to-cut materials, nevertheless it is hardly obtained for milling operations.
- Plastic deformation, untempered martensite (white layer) and overtempered martensite (dark layer) are the microstructure alterations most frequently observed after subjecting hardened steels to cutting by shearing and grinding. The occurrence of plastic deformation and white layer is closely related to the cutting-edge preparation and tool wear and tend to be more intense when tools with large hone radius or severely worn tools are employed. Cracks, recast layer and heat-affected zone are microstructure alterations frequently observed after EDM of hardened steels.
- The microhardness profile along the depth below the machined surface is affected somewhat by the measuring method (indenter geometry and applied load). The availability of equipment enabling the use of loads in the range of millinewtons (nanohardness measurement) has allowed a more accurate assessment of the hardness profile. The hardness behaviour usually observed in hardened steels consists of a peak value on the surface (suggesting the presence of an untempered martensite layer), followed by a steep decrease to values below the bulk hardness (overtempered martensite layer) before returning to the original hardness value.
- The resulting residual stress of the machined component depends on a combination of mechanical (plastic deformation) and thermal effects (microstructure changes) that take place during machining. As far as the operation is concerned, abusive grinding and EDM tend to generate tensile residual stresses on hard-

ened steel components, whereas turning, milling, gentle grinding and high-speed machining generally induce compressive stresses. Cutting tools possessing a honed edge tend to induce compressive stresses; however, the influence of the cutting parameters and tool wear is not consensual. In general, the published literature suggests that increasing cutting speed results in tensile stresses, whereas the elevation of feed rate promotes compressive stresses and the influence of depth of cut is regarded as negligible.

- The fatigue resistance of a machined component depends on both surface finish and residual stress induced by the operation: the better the surface finish and the more compressive the residual stress, the longer the fatigue life. Tool nose radius seems to be the principal parameter affecting fatigue life and, interestingly, best results are obtained when the fatigue specimens are produced using tools with small nose radius.

Acknowledgements The authors would like to thank Cláudio Henrique Dias de Almeida, Marcelo da Mota Silva, Steve Balbino Diniz and Wagner Campos for their support to part of the experimental work.

References

[1] König W, Berktold A, Koch K-F (1993) Turning versus grinding – a comparison of surface integrity aspects and attainable accuracies. Ann CIRP 42:39–43

[2] Rech J, Moisan A (2003) Surface integrity in finish hard turning of case-hardened steels. Int J Mach Tools Manuf 43(5):543–550

[3] Abrão AM, Aspinwall DK, Wise MHL (1995) Tool life and workpiece surface integrity evaluations when machining hardened AISI H13 and AISI E52100 steels with conventional ceramic and PCBN tool materials. SME technical paper. Society of Manufacturing Engineers MR95-159, Dearborn, MI, pp 1–7

[4] Javidi A, Rieger U, Eichlseder W (2008) The effect of machining on the surface integrity and fatigue life. Int J Fatigue 30(10–11):2050–2055

[5] Jacobson M, Dahlman P, Gunnberg F (2002) Cutting speed influence on surface integrity of hard turned bainite steel. J Mater Process Technol 128:318–323

[6] Ávila RF, Abrão AM (2001) The effect of cutting fluids on the machining of hardened AISI 4340 steel. J Mater Process Technol 119:21–26

[7] Benga GC, Abrão AM (2003) Turning of hardened 100Cr6 bearing steel with ceramic and PCBN cutting tools. J Mater Process Technol 143:237–241

[8] Lima JG, Ávila RF, Abrão AM, Faustino M., Davim JP (2005) Hard turning: AISI 4340 high strength low alloy steel and AISI D2 cold work tool steel. J Mater Process Technol 169:388–395

[9] Davim JP, Figueira L (2007) Comparative evaluation of conventional and wiper ceramic tools on cutting forces, surface roughness, and tool wear in hard turning AISI D2 steel. Proc Int Mech Eng B: J Eng Manuf 221:625–633

[10] Gaitonde VN, Karnik SR, Figueira L, Davim JP (2009) Machinability investigations in hard turning of AISI D2 cold work tool steel with conventional and wiper ceramic inserts. Int J Refract Met Hard Mater 27:754–763

[11] Koshy P, Dewes RC, Aspinwall DK (2002) High speed end milling of hardened AISI D2 tool steel (~58 HRC). J Mater Process Technol 127:266–273

[12] Elbestawi MA, Chen L, Becze CE, El-Wardany TI (1997) High-speed milling of dies and molds in their hardened state. Ann CIRP 46(1):57–62
[13] Coldwell H, Woods R, Paul M, Koshy P, Dewes R, Aspinwall D (2003) Rapid machining of hardened AISI H13 and D2 moulds, dies and press tools. J Mater Process Technol 135:301–311
[14] Waikar RA, Guo YB (2008) A comprehensive characterization of 3D surface topography induced by hard turning versus grinding. J Mater Process Technol 197(1–3):189–199
[15] Toh CK (2004) Surface topography analysis in high speed finish milling inclined hardened steel. Precision Engineering 28:286–398
[16] Sharman ARC, Hughes JI, Ridgway K (2008) Surface integrity and tool life when turning Inconel 718 using ultra-high pressure and flood coolant systems. Proc Inst Mech Eng Part B: J Eng Manuf 222:653–664
[17] Coelho RT, Silva LR, Braghini Jr A, Bezerra AA (2004) Some effects of cutting edge preparation and geometric modifications when turning INCONEL 718™ at high cutting speeds. J Mater Process Technol 148:147–153
[18] Arunachalam RM, Mannan MA, Spowage AC (2004) Surface integrity when machining age hardened Inconel 718 with coated carbide cutting tools. Int J Mach Tools Manuf 44:1481–1491
[19] Sharman ARC, Amarasinghe A, Ridgway K (2008) Tool life and surface integrity aspects when drilling and hole making in Inconel 718. J Mater Process Technol 200:424–432
[20] Matsumoto Y, Hashimoto F, Lahoti G (1999) Surface integrity generated by precision hard turning. Ann CIRP 48(1):59–62
[21] Field M, Kahles JF, Koster WP (1989) Surface finish and surface integrity. In: Metals handbook, vol 16, Machining, 9th edn. ASM International Materials Park, pp 19–36
[22] Bosheh SS, Mativenga PT (2006) White layer formation in hard turning of H13 tool steel at high cutting speeds using CBN tooling. Int J Mach Tools Manuf 46:225–233
[23] Axinte DA, Dewes RC (2002) Surface integrity of hot work tool steel after high speed milling – experimental data and empirical models. J Mater Process Technol 127:325–335
[24] Guo YB, Sahni J, (2004) A comparative study of hard turned and cylindrically ground white layers. Int J Mach Tools Manuf 44:135–145
[25] García Navas VG, Ferreres I, Marañón JA, García-Rosales C, Gíl Sevillano J (2008) White layers generated in AISI O1 tool steel by hard turning or by EDM. Int J Mach Mach Mater 4(4):287–301
[26] Thiele JD, Melkote SN (2000) Effect of tool edge geometry on workpiece subsurface deformation and through-thickness residual stresses for hard turning of AISI 52100 steel. J Manuf Process 2(4):270–276
[27] Smith S, Melkote SN, Lara-Curzio E, Watkins TR, Allard L, Riester L (2007) Effect of surface integrity of hard turned AISI 52100 steel on fatigue performance. Mater Sci Eng A 459(1–2):337–346
[28] Abrão AM, Aspinwall DK (1996) The surface integrity of turned and ground hardened bearing steel. Wear 196:279–284
[29] Matsumoto Y, Magda D, Hoeppner DW, Kim TY (1991) Effect of machining processes on the fatigue strength of hardened AISI 4340 steel. J Eng Ind 113:154–159
[30] Tönshoff HK, Brinksmeier E (1980) Determination of the mechanical and thermal influences in machined surfaces by microhardness and residual stress analysis. Ann CIRP 29(2):519–529
[31] Warren AW, Guo YB, Weaver ML (2006) The influence of machining induced residual stress and phase transformation on the measurement of subsurface mechanical behavior using nanoindentation. Surf Coat Technol 200(11):3459–3467
[32] Hashimoto F, Guo YB, Warren AW (2006) Surface integrity difference between hard turned and ground surfaces and its impact on fatigue life. CIRP Ann Manuf Technol 55(1):81–84
[33] Pawade RS, Joshi SS, Brahmankar PK (2008) Effect of machining paramenters and cutting edge geometry on surface integrity of high-speed turned Inconel 718. Int J Mach Tools Manuf 48:15–28

[34] Matsumoto Y, Barash MM, Liu CR (1986) Effect of hardness on the surface integrity of AISI 4340 steel. J Eng Ind 108:169–175
[35] Novovic D, Dewes RC, Aspinwall DK, Voice W, Bowen P (2004) The effect of machined topography and integrity on fatigue life. Int J Mach Tools Manuf 44:125–134
[36] Choi Y (2009) A study on the effects of machining-induced residual stress on rolling contact fatigue. Int J Fatigue 31:1517–1523
[37] Gunnberg F, Escursell M, Jacobson M (2006) The influence of cutting parameters on residual stresses and surface topography during hard turning of 18MnCr5 case carburised steel. J Mater Process Technol 174(1–3):82–90
[38] Dahlman P, Gunnberg F, Jacobson M (2004) The influence of rake angle, cutting feed and cutting depth on residual stresses in hard turning. J Mater Process Technol 147:181–184
[39] Xueping Z, Erwei G, Liu CR (2009) Optimization of the process parameter of residual stresses for hard turned surfaces. J Mater Process Technol 209:4286–4291
[40] Liu M, Takagi J, Tsukuda A (2004) Effect of tool nose radius and tool wear on residual stress distribution in hard turning of bearing steel. J Mater Process Technol 150:234–241
[41] Arunachalam RM, Mannan MA, Spowage AC (2004b) Residual stress and surface roughness when facing age hardened Inconel 718 with CBN and ceramic cutting tools. Int J Mach Tools Manuf 44:879–887
[42] Li W, Withers PJ, Axinte D, Preuss M, Andrews P (2009) Residual stresses in face finish turning of high strength nickel-based superalloy. J Mater Process Technol 209:4896–4902
[43] Segawa T, Sasahara H, Tsutsumi M (2004) Development of a new tool to generate compressive residual stress within a machined surface. Int J Mach Tools Manuf 44:1215–1221
[44] Boardman B (1990) Fatigue resistance of steels. In: Metals handbook, vol 1, Properties and selection: iron, steels and high performance alloys, 10th edn. ASM International Materials Park, pp 673–688
[45] Czyryca EJ (1989) Fatigue crack initiation. In: Metals handbook, vol 8, Mechanical testing, 9th edn. ASM International Materials Park, pp 366–375
[46] Guo, YB, Warren AW, Hashimoto, F (2010) The basic relationships between residual stress, white layer, and fatigue life of hard turned and ground surfaces in rolling contact. CIRP J Manuf Sci Technol (in press)

Chapter 5
Finite-element Modeling and Simulation

P.J. Arrazola

This chapter deals with the finite-element method (FEM) of hard machining, mainly turning (two- and three-dimensional (3D)). Results about the influence of working conditions and tool geometry (cutting-edge finishing) on tool forces, temperatures, and stresses when machining AISI 52100 steel are presented. In addition, information about residual stresses obtained through 3D FEM analysis is shown. The aim of the chapter is to demonstrate the possibilities of FEM for understanding the chip formation process in hard turning and to show its capabilities in areas like tool insert design and prediction of the surface state of the machined workpiece. First, a brief summary of the state of the art on hard machining is presented. Then FEM capabilities and limitations are shown. After that, results of process simulations will be provided and compared with those obtained in the literature. Finally, overall conclusions are pointed out and future research direction is discussed.

5.1 Introduction

Hard turning has become a relevant manufacturing process in producing finished components that are made of alloyed steels with hardnesses between 50 and 70 HRC [1]. The main aim of employing hard turning is avoiding the grinding operation which in most cases corresponds to the final operation for the workpiece. Comparisons between both processes show that the demanded surface roughness and ISO tolerance standards can be achieved in both processes, but higher flexibility and material removal rates can be obtained in hard turning.

P.J. Arrazola
Faculty of Engineering, Mondragon University, Loramendi 4; 20.500-Mondragón, Spain
e-mail: pjarrazola@eps.mondragon.edu

Moreover, minimum quantity of lubricant or even dry machining can be employed in hard turning [1–3].

In hard turning, surface integrity becomes a relevant customer requirement. Thus, issues like surface roughness, residual stresses and material microstructure, tool behavior or tool life are often carefully studied [4–7].

Hard turning and grinding processes create different surface microgeometry [8]. With an increasing surface roughness, the influences of the microgeometrical valleys obtained in turning acquire more relevance with regard to fatigue life.

In hard turning, thermal and mechanical loads are quite significant. If the temperature promoted by thermal loads exceeds the austenite formation temperature, changes in the material's microstructure can occur. These structural changes mainly depend on the heating and cooling rates as well as the maximum temperature reached in the contact area [1, 9]. This aspect affects the physical properties of the workpiece surface and thus the fatigue life of the component [2–7, 10].

White layers in the workpiece surface are produced by both grinding and turning processes, and are considered detrimental to component performance, primarily in relation to fatigue [12–14]. As far as the white layer is concerned, from the literature three different theories explaining the structure of white layer formation have emerged:

- rapid heating and quenching, which results in sudden microstructural transformation;
- severe plastic deformation, which produces a homogenous structure and/or a microstructure producing a very fine grain size;
- surface reaction with the environment, such as in nitriding processes.

Some authors mention the existence of dark layers which seem to be a result of microstructural changes in the heat-affected zone as a consequence of the rapid heating and quenching [15].

As for residual stresses, in hard turning the maximum tensile stress occurs at the surface while grinding; it is usually located underneath [1, 3, 4]. Nevertheless, the affected zone seems to be higher in grinding. Apart from that, the influence of microstructure changes can affect in a more remarkable way the residual-stress distribution to fatigue life.

Smith *et al.* [16] analyzed fatigue results for turned and ground AISI 52100 steel (60–62 HRC). It was observed that hard-turned specimens exhibited at least as high a fatigue life as the baseline ground specimens. It was concluded as well that the effect of residual stress on fatigue life is more significant than the effect of white layers.

Abrasion, diffusion, and attrition are the most common tool wear mechanisms [18]. Although tools with ceramic inserts are employed, the most common tool material employed in hard turning is polycrystalline cubic boron nitrite (PCBN). There are several types of PCBN inserts depending on the percentage of PCBN, the grain size, and the binder. For instance, low-content PCBN seems to be the appropriate choice for hard turning.

Cutting edges of PCBN inserts are especially prepared with a honed finish or even a chamfer in order to better withstand the high stresses that appear in this area during the chip formation process. This microgeometry of the cutting edge plays a key role in the workpiece surface properties and the performance of the cutting tool [18]. The design of cutting edges may affect the chip formation mechanism and therefore help in reducing cutting forces and increasing tool life. Therefore, tool manufacturers introduce different types of tool edge preparations such as chamfer, double chamfer, chamfer plus hone, hone, and oval or parabolic edge designs [19, 20].

Due to the material hardness, the typical saw-tooth or serrated chip is obtained during the chip formation process [21].

The hard-turning operation is usually carried out at cutting speeds of 100–200 m/min, feed rates of 0.1–0.2 mm and depth of cut of less than 0.5 mm. The operation should be carried out with machine tools that have high stiffness and damping capacity in order to obtain the expected tool life values.

5.2 Finite-element Modeling

The finite-element method can provide a comprehensive and in some cases complementary approach to experimental, mechanistic or analytical approaches to study machining process [19, 23–25]. It offers capability to predict what could happen during the material removal process, and thus it could be possible to design and modify the process input parameters beforehand in order to reduce or eliminate problems that may arise during actual machining operations.

Since 1970s, there have been several applications of finite-element modeling (FEM) in chip formation processes [26–28]. Firstly, custom-made codes and later, basically with general-purpose software like Abaqus™, NIKE (forging), ALGOR™ (metal forming), FORGE2D™, DEFORM-2D™, MARC™, LS DYNA™, and FLUENT™ [28]. In addition, two specific commercial software programs have appeared on the market during the last decade: AdvantEdge™ and DEFORM™.

Unlike earlier chip formation cutting models [30–34], FEM of chip formation processes [35–37] provides some advantages, basically due to the following aspects: access to fields of values of thermo-mechanical variables, consideration of the nonlinear effects of the friction at the tool–chip interface, ability to perform virtual machining tests that are difficult to justify experimentally (new tool geometries and materials or coatings) or to study cases difficult to carry out (zero friction coefficient, materials with ideal behavior, *etc.*).

It can be said that basically three kinds of problems can be found related to numerical cutting modeling, prior to becoming a reliable tool for industry:

- Finite-element model definition: boundary conditions, integration frame (explicit, implicit), formulations can lead to different quantitative results, calculation times, *etc.*

- Finite-element model validation: except for parameters like forces, chip thickness or tool–chip contact length, experimental measurement is quite complicated (*e.g.*, temperature, strain, strain rate) and in some cases quite difficult at this stage (strain rate, stresses, *etc.*). Thus model validation is quite difficult because experimental means are not available or they are not suitable enough.
- Identification of finite-element model input parameters.

A sensitivity study showed the influence of input parameters on results [38]. From this study, information about the uncertainty originated by input parameter identification can be estimated and the remarkable difficulties in obtaining quantitative results can be pointed out [38].

For instance, moving the yield stress (one of the best-known material coefficients) from 200 MPa to 900 MPa can lead to a temperature increase of 30 % from an average value of 1240 K (*i.e.*, of around 372 K). Thus it can be estimated that an uncertainty of 30 MPa can give an uncertainty in temperatures of approximately 15 K [39].

The thermo-mechanical model of the chip formation process only takes into account one area of the part and the tool, the area where the chip is formed. In most of the research works done until now, instead of a three-dimensional (3D) model, a two-dimensional (2D) model is considered, that is say orthogonal cutting, where it is supposed to be in a plane stress case. Even if this approach is a restriction from a machining point of view (often a 3D case), it is considered accurate enough in many cases to give an overview of what is happening. Furthermore, it allows a reduction of computational time.

However, it should be clear that in 2D modeling of many machining operations: turning, drilling, milling, *etc.* (except broaching, sawing, *etc.*) the obtained final surface doesn't correspond to the final one that is obtained in 3D. Thus, some prediction issues like residual stresses cannot have any realistic meaning.

According to the formulation type, it is possible to establish a classification: Lagrangian, Eulerian or arbitrary Eulerian Lagrangian (ALE).

In the Lagrangian description the domain is discretized with a mesh that is going to follow the material at every moment, consequently nodes move simultaneously with this last one. The main problem is the mesh control due to the remarkable distortions that appear during the chip formation. Two solutions are generally used:

- The uncut chip portion of the workpiece is "glued" over the workpiece final surface: the chip is separated from the workpiece when a geometric or mechanical criterion is reached [40–45].
- Remeshing: when the distortions of the mesh of the workpiece are extremely high, a new mesh is generated [19–21, 24, 25, 37, 45–49]. After a mapping of the data from the distorted mesh to the new one, the calculation continues until the criterion that determines the regeneration of the mesh is reached again. The criterion that promotes the mesh renovation can be, for example, the plastic strain.

The Eulerian description considers a fixed region in the geometric space occupied by continuum material examined over time. The space domain is discretized and the movement must be determined with a mesh that stays fixed at every moment. That is, the nodes do not move during the analysis and they are not fixed to the material [9, 35, 36, 50–52]. This approach is very fast to execute, but great difficulties appear when dealing with the free-surface treatment. Indeed, as in the beginning of the simulation the chip shape is not known, adaptation algorithms of the mesh will be needed.

As in the case of the Lagrangian formulation, to predict the serrated chip formation, a crack propagation criterion is needed. On the other hand, this Eulerian approach is used solely in the steady-state analysis.

Finally, another method of description of the movement is the ALE or mixed description [46, 53–58]. In this formulation, the nodes can move, but always remain inside the boundary region defined in the ALE domain. The major advantages are: easy application of the boundary conditions, easy treatment of the interfaces and lower distortion of the mesh, *etc.*

5.2.1 Commercial Software

The two major commercial software programs existing on the market are AdvantEdge and DEFORM-2D/-3D. Both programs provide an interface to the end user, in order to ease the introduction of process parameters and in a way make transparent those subjects dealing with the mathematical theory of the finite-element method.

DEFORM is software specialized for modeling machining operations in 2D and 3D based on an implicit integration method, with fully coupled thermo-mechanical analysis. Since the formulation is Lagrangian, an adaptive remeshing technique is used to reduce the mesh distortions when the chip is formed.

Several material constitutive laws can be used to model the material behavior (Oxley's equation, Johnson–Cook equation, among others). A material database can be found in this software, both for tool and workpiece materials. For modeling the contact at the tool–chip interface, a constant shear factor friction law or Coulomb friction law can be employed.

Workpiece and tool geometries should be configured by the user, both in terms of the external dimensions and those of the mesh of the two parts. However, coating layers can be implemented in the tool. Also remarkable is the availability of loading some existing geometries of tools and toolholders from a database incorporated in the program.

AdvantEdge is an explicit dynamic, fully thermo-mechanically coupled finite-element program specialized in modeling machining operations in 2D and 3D.

Lagrangian techniques are employed for modeling the metal-cutting process. Tracking discrete material points and using a predetermined line of separation to generate a fictitious crack, is possible to obtain the chip and the workpiece. As the formulation is Lagrangian, an adaptive remeshing technique is used to reduce the mesh distortions.

The material constitutive law employed is the Marusich–Ortiz law [44]. Other material models can be implemented through the use of subroutines. As in DEFORM, a material database can be found in this software, both for tool and workpiece materials. For modeling the contact at the tool–chip interface the Coulomb friction law is employed.

Specific process characteristics can be defined like the use of coolant and its properties, the friction coefficient at the contact or the option to predict burrs in the workpiece.

5.2.2 State of the Art in Finite-element Models of Hard Turning

In order to reduce the experimental costs, FEM of hard turning can be employed to qualitatively predict tool forces, stress, temperature, strain and strain rate fields. Having this in mind, with the existing difficulties to identify input parameters properly it is not possible to obtain reliable quantitative values. It also allows the estimation of the superficial state of the machined surface, that is, the residual stresses induced in the workpiece. A finite-element model for 3D turning has also been introduced in earlier studies. Guo and Liu [59] proposed 3D FEM for hard turning of AISI 52100 steel using PCBN tools. The model was used to predict the cutting forces, temperature distribution over the cutting edge, and the residual-stress distribution on the machined surface. Effects of edge preparation in cutting tools for hard turning have been studied using FEM simulations by Yen *et al.* [60] and Chen *et al.* [59]. Klocke and Kratz [18] used 3D FEM to calculate the temperature on a chamfered-edge PCBN tool design. Aurich and Bil [62] proposed a 3D finite-element model for predicting serrated chip formation when machining Inconel 718. Özel *et al.* [18] proposed a 3D model to study the influence of PCBN tool microgeometry. Arrazola and Özel [54] proposed a 3D finite-element model for predicting residual stresses developed with Abaqus/Explicit™. Attanasio *et al.* [25] proposed a 3D finite-element model for prediction of tool wear in metal-cutting operations when turning AISI 1045 steel using uncoated tungsten carbide tools. Kountanya *et al.* analyzed experimentally and with finite-element simulations the influence of cutting conditions and tool edge geometry when machining AISI-E52100 steel (in 2D) [63]. Umbrello and Filice [64] proposed a 2D model employing a hardness-based flow stress and an empirical model for describing the white- and dark-layer formation. They also proposed the development of hardness-based flow stress and fracture models for machining AISI H13 tool steel, which could be applied for a wide range of work material hardness [63].

It is worthwhile mentioning that during recent years, several other research works have been developed in 3D [8, 19, 24, 25, 66] showing the remarkable future possibilities of FEM of chip formation processes.

5.3 Finite-element Modeling of Hard Turning

In order to show qualitatively the possibilities of FEM applied to hard turning, in the following paragraphs some simulations will be carried out in 2D and 3D using the commercial software AdvantEdge and DEFORM-2D, and the general-purpose finite-element analysis software Abaqus/Explicit (v6.6-1). Results about forces, temperatures, stresses, pressures, and residual stresses (only in 3D analysis) will be provided and compared qualitatively to experimental results found in the literature.

5.3.1 Two-dimensional Finite-element Analysis of Hard Turning

Two 2D finite-element models have been employed to study the influence of cutting parameters (cutting speed, feed rate, and cutting-edge finishing) in hard turning: AdvantEdge and general-purpose FEM software Abaqus/Explicit (v6.6-1). Information about the model developed with Abaqus/Explicit is presented in Table 5.1. Data about AdvantEdge correspond to the closest to Abaqus/Explicit found in the software database.

Twelve simulations were carried out with AdvantEdge, varying the cutting speed (v) from 100 to 140 m/min, and the feed rate (f) from 0.1 mm/rev to 0.15 mm/rev. Two types of cutting-edge finishing have been employed: (1) a cutting-edge radius of 25 μm and (2) a chamfer of 20° × 0.1 mm. The whole experimental plan and results are shown in Table 5.2.

The material employed in these calculations is defined by the program database (AISI 52100), and its law cannot be modified by the user. Only material hardness can be modified in AdvantEdge and it has been set to 62 HRC.

In the case of Abaqus/Explicit only one simulation has been carried out, aiming to have a reference point for comparison with AdvantEdge results. Although the limitations have been reported in the literature [67] the thermo-visco-plastic behavior of the workpiece is modeled by the constitutive Johnson-Cook equation [68]:

$$\bar{\sigma} = \left[A + B(\bar{\varepsilon})^n\right]\left[1 + C\ln\left(\dot{\bar{\varepsilon}}/\dot{\bar{\varepsilon}}_0\right)\right]\left[1 - \left(\frac{\theta_W - \theta_0}{\theta_m - \theta_0}\right)^m\right] \tag{5.1}$$

where $\bar{\varepsilon}$ is the plastic strain, $\dot{\bar{\varepsilon}}$ is the plastic strain rate, $\bar{\varepsilon}_0$ is the reference plastic strain rate (0.001 s^{-1}), θ_w is the temperature of the workpiece, θ_m is the melting temperature of the workpiece material, and θ_0 (293 K) is the room temperature. Material constant A is the yield strength, B is the hardening modulus, C is the strain rate sensitivity, n is the strain-hardening exponent, and m the thermal softening exponent. Although a more realistic simulation model for the machining process should also take into account the state of the work material due to a previous machining pass or manufacturing process, in our model the material enters the workpiece without any strain or stress history.

The Coulomb friction law has been employed for the modeling of the tool–chip interface contact. Heat transfer is allowed on the tool–chip contact area.

The elements employed are CPE4RT, temperature calculation, and hourglass control.

Table 5.1 Material properties and cutting conditions for the process simulations (Abaqus/Explicit and DEFORM-3D)

			Abaqus/Explicit, DEFORM-3D
Material properties	Plasticity, Johnson–Cook law	A (MPa)	2482.4
		B (MPa)	1498.5
		n	0.19
		C	0.027
		m	0.66
	Inelastic heat fraction (β)		0.9
	Density (ρ) (kg m^{-3})	Workpiece (AISI 52100 62HRC)	7827
		Tool (PCBN)	3120
	Elasticity (E) (GPa)	Workpiece	201.33 f(T)
		Tool	680
	Conductivity (k) (W m^{-1} K^{-1})	Workpiece	43
		Tool	100
	Specific heat (c) (J kg^{-1} K^{-1})	Workpiece	458
		Tool	960
	Expansion (K^{-1})	Workpiece	11.5×10^{-6} f(T)
		Tool	4.9×10^{-6}
Contact	Thermal conductance (K_i) (W m^{-2} K^{-1})		1×10^{8}
	Heat partition coefficient (Γ)		0.5
	Friction coefficient (μ)		0.35 (Guo)
	Friction energy transformed into heat (η)		1
Process	Cutting speed (v) (m/min)		100/120/140
	Feed rate (f) (mm/rev)		0.1/0.15
	Cutting-edge microgeometry		Chamfer 20° × 0.1 mm
	Rake angle (γ) (°)		–5
	Clearance angle (α) (°)		5

Table 5.2 Experimental plan and results obtained for Abaqus/Explicit and AdvantEdge

Abaqus/Explicit, AdvantEdge	Cutting Forces		Tool	Workpiece			Chip			Shear plane stress
	F_c (N)	F_f (N)	Temp (K)	Temp (K)	ε	σ (GPa)	Temp (K)	ε	σ (GPa)	(GPa)
Vc_100_Av_010_r25	63	50	943	769	3	0.5	957	18.4	0.11	3.6
Vc_120_Av_010_r25	58	49	987	793	3	0.66	991	16.6	0.12	3.03
Vc_140_Av_010_r25	56	47	1013	768	2.8	0.82	1015	16.4	0.132	3.3
Vc_100_Av_015_r25	86	55	958	843	4.1	0.45	979	22	0.08	3.1
Vc_120_Av_015_r25	82	52	1106	902	6.1	0.3	1111	26	0.1	3.2
Vc_140_Av_015_r25	80	53	1080	853	4.1	0.72	1087	27	0.068	3.1
Vc_100_Av_010_ch	72	54	1065	878	4	0.3	1077	18.8	0.07	2.6
Vc_120_Av_010_ch	69	54	1079	978	4	0.1	1083	19	0.12	2.1
Vc_140_Av_010_ch	68	55	1093	983	5	0.12	1094	18	0.2	2.8
Vc_100_Av_015_ch	95	69	1136	973	10	0.17	1138	26	0.158	3.4
Vc_120_Av_015_ch	94	68	1143	1123	9.5	0.28	1081	18	0.151	2.9
Vc_140_Av_015_ch	94	80	1178	1043	10	0.2	1179	30	0.06	2.5
Vc_100_Av_010_r25 (Abaqus)	90	60	1350	1150	8	1	1400	35	1.5	3.9

Figure 5.1 shows the zones where the variable results are extracted. As can be seen, the data about chips is extracted at the secondary shear zone, where the highest temperature and plastic strains occur. In the case of the workpiece, the tertiary zone has been selected for extracting data.

The element size employed in the AdvantEdge model is around 5 μm, the same as the Abaqus model. The minimum and maximum element sizes vary significantly, especially in the AdvantEdge model. The maximum value obtained in AdvantEdge is 75 μm, while in Abaqus the biggest elements are about 30 μm.

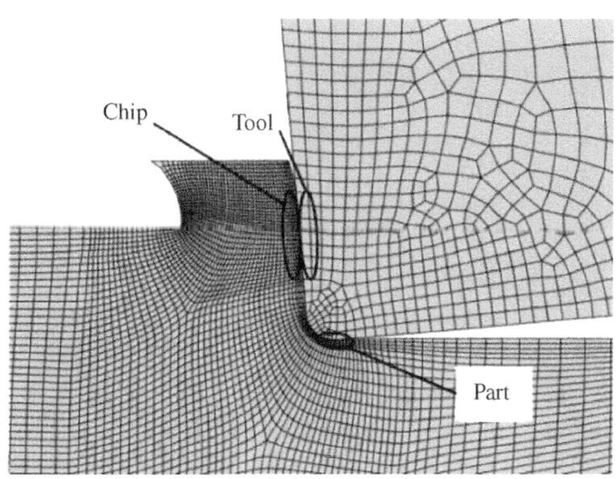

Figure 5.1 Results extraction zones

5.3.2 *Two-dimensional Finite-element Analysis of Hard Turning: Results and Discussion*

Simulations were run until the steady state was reached, in the case of Abaqus/Explicit at 1.6 ms, and 0.75 ms in the case of AdvantEdge. In the case of Abaqus, a previous stabilization process was carried out in order to reach the required cutting speed, in this case 100 m min^{-1}. As a consequence of this procedure, from the 1.6 ms stated before only 1 ms is carried out at the specified cutting speed, while the rest of the time cutting is at lower speeds.

As seen in Figure 5.2 (AdvantEdge), cutting force seems to stabilize around 70 N, while temperature (Figure 5.3, AdvantEdge) stabilizes at 1000 K after approximately 0.17 ms.

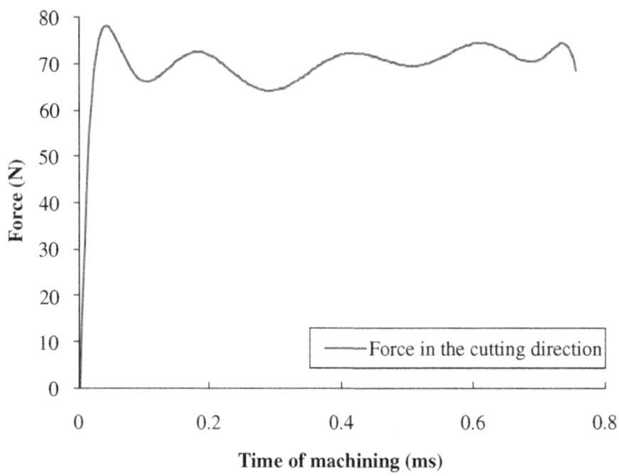

Figure 5.2 Cutting force in AdvantEdge

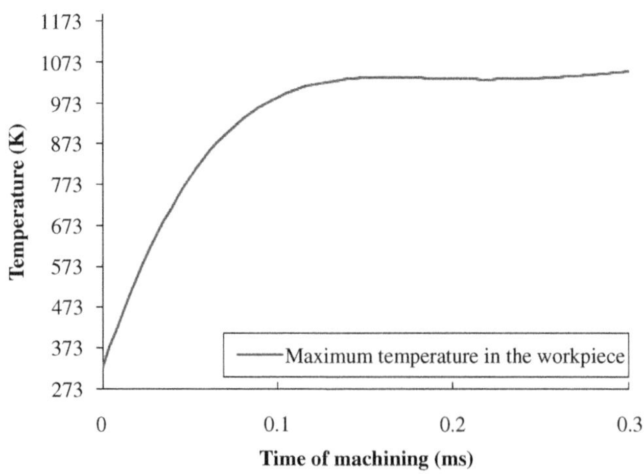

Figure 5.3 Workpiece maximum temperature in AdvantEdge

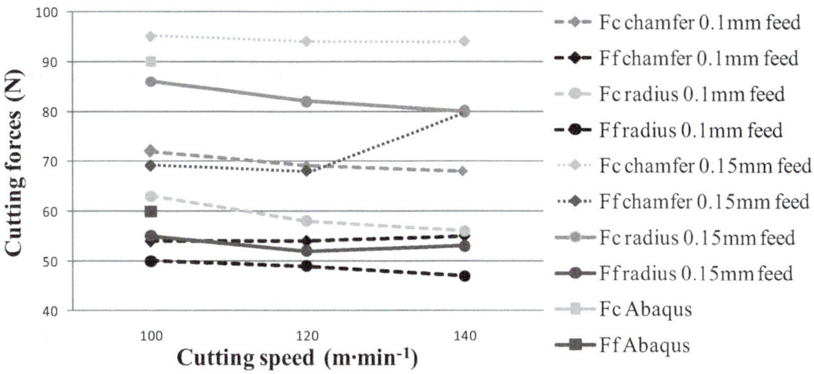

Figure 5.4 Cutting forces obtained in all the tests

Figure 5.4 shows the cutting and feed forces depending on the cutting speed, the undeformed chip thickness (feed rate) and the cutting-edge finishing.

Comparing cutting forces, fairly different results have been obtained in the two programs, with cutting forces around 90 N in Abaqus and 60 N in AdvantEdge. The feed forces match better and vary from 60 to 50 N.

Comparing the results obtained with AdvantEdge it is observed that with the chamfer finishing cutting and feed forces are higher than with a round edge finishing. For instance at the cutting speed of 120 m/min and feed rate of 0.15 mm the cutting force increases from 82 to 94 N. As for the feed force, it increases from 52 to 69 N.

As for the feed rate, it is also observed that moving the feed rate from 0.1 to 0.15 mm induces an increase in the cutting and feed forces in the same proportion. For instance at the cutting speed of 120 m/min and with a round chamfer finishing edge, the cutting force increases from 69 to 94 N and the feed force from 53 to 69 N.

The influence of the cutting speed seems to be more negligible with a slight decrease of both cutting and feed forces as the cutting speed increases.

Figure 5.5 shows the temperature fields for both radius-tooled models at 100 m min^{-1}. A maximum temperature of 1350 K is achieved in the tool in

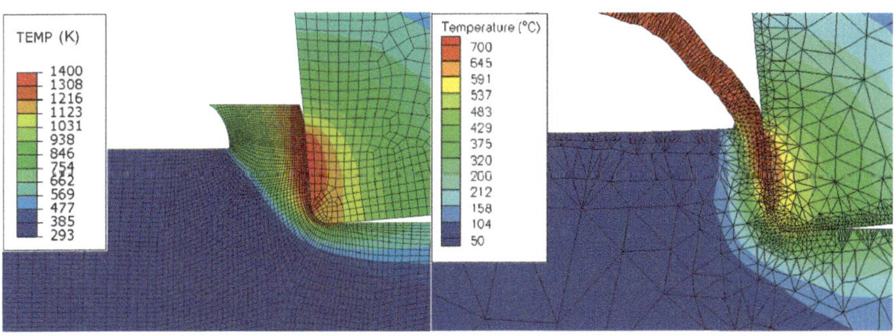

(a) (b)

Figure 5.5 Temperature fields in Abaqus (a), and AdvantEdge (b). Cutting speed of 100 m min^{-1}, feed rate of 0.1 mm, and cutting radius of 25 μm

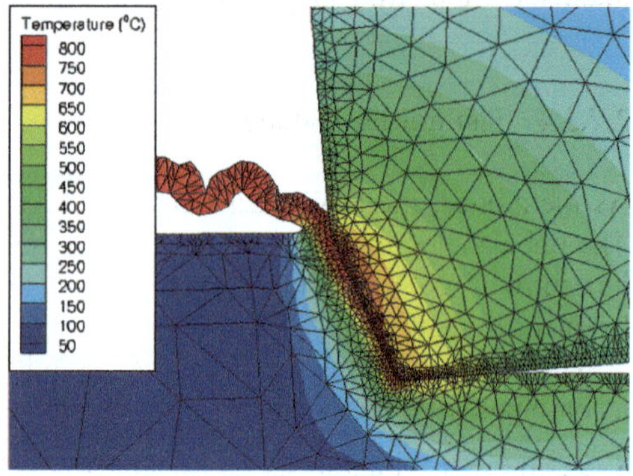

Figure 5.6 AdvantEdge: cutting speed of 100 m min^{-1}, feed rate of 0.1 mm, and chamfered cutting edge

Abaqus, while the maximum temperature obtained in AdvantEdge is 958 K. These differences can be expected as the input parameters are not the same, although it was attempted to have similar ones.

In the case of the workpiece, Abaqus reaches 1150 K, while in AdvantEdge the maximum temperature is 843 K.

Figure 5.6 shows the temperature fields for the hard-turning simulation in the tool and workpiece for the chamfered model at 120 m min^{-1}. A temperature of 1079 K is achieved in the tool, while the maximum temperature in the workpiece reaches 978 K. Compared to the round-edge model (Figure 5.5 (b)), higher temperatures have been obtained both in the tool and workpiece. It is also observed that higher thermal-affected zones are obtained when employing chamfered cutting edges.

Figure 5.7 shows the temperature trends in the tool and workpiece depending on the cutting speed, cutting-edge geometry, and feed. As the tool and chip temperatures are very similar, only tool values are shown. In general it is observed that as the cutting speed increases there is a slight increase in temperature values.

In all the cases the maximum temperature is reached in the tool. Comparing the round edge with the chamfered edge, it is observed that higher values are obtained for the latter (as mentioned before). For the chamfered tool, a maximum of 1079 K is attained in the tool and 978 K in the workpiece. Using round cutting edges seems to decrease the maximum temperature, achieving 987 K in the tool

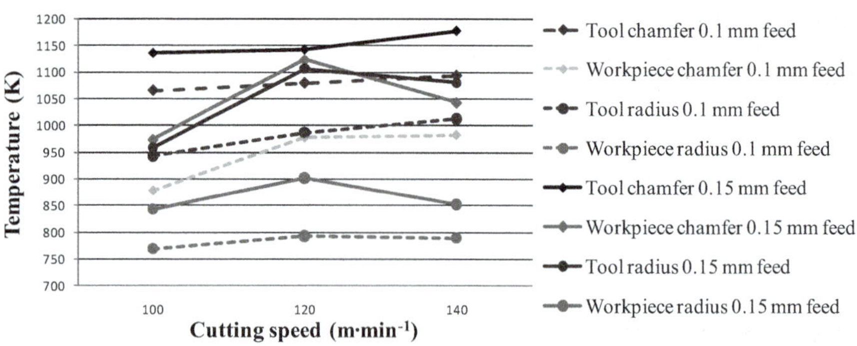

Figure 5.7 Cutting speed, feed and cutting-edge finishing influence in tool temperatures

and, remarkably, only 793 K in the part – almost 200 K below that in the chamfered-edge model. Thus from the thermal point of view, round cutting edges would seem to withstand hard machining better.

As for the feed rate, when comparing values for a feed rate of 0.1 mm and 0.15 mm differences of roughly 100 K are found (lower in the former).

Moving the cutting speed from 100 to 140 m min^{-1} causes a slight rise in the tool temperature, around 50 K. Nevertheless, for the 0.15-mm-feed models a singular effect has been observed: the workpiece temperatures suffer a high increase at the cutting speed of 120 m min^{-1}.

Figure 5.8 shows the plastic-strain distribution in Abaqus and AdvantEdge. Extremely high values, close to 20, are found in both models. It can be seen in Figure 5.9 that similar maximum plastic-strain values are found in the chamfered-edge and round-edge models (close to 9–10). The plastically affected zone in the radius-tooled model is 20 μm in depth, while in the chamfered model this zone extends to about 600 μm. These higher values in the workpiece plastically affected zone can be the cause of the temperature differences in the machined surface and may affect the residual stresses.

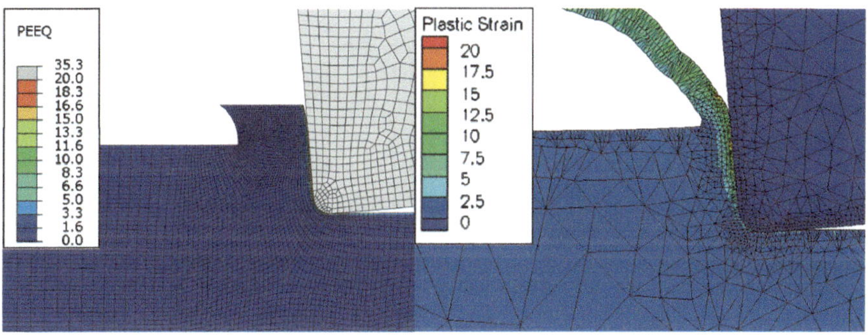

(a) (b)

Figure 5.8 Plastic strain in Abaqus (a) and AdvantEdge (b). Cutting speed of 100 m min^{-1}, feed rate of 0.1 mm and cutting radius of 25 μm

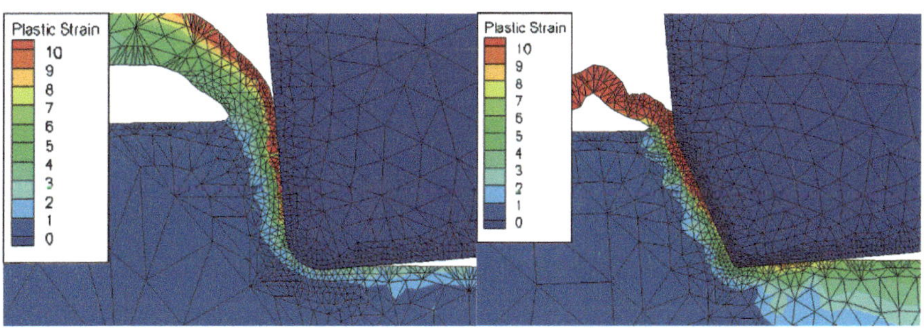

(a) (b)

Figure 5.9 Plastic equivalent deformation plots of radius model (a) and chamfered model (b) in AdvantEdge at 100 m min^{-1} and 0.1 mm feed

Figure 5.10 shows the stress distribution in Abaqus and AdvantEdge. Similar stress values are obtained in the case of Abaqus compared with AdvantEdge for the workpiece material (3.9 GPa).

Nevertheless, regarding the stress fields obtained, in Figure 5.11 it is clearly visible how higher maximum stresses are obtained in the round-finished tool compared with the chamfered tool (4 GPa for round cutting edge and 2.5 GPa for chamfered tool). This means that for the same cutting conditions lower stresses will be found in the chamfered tool, and longer tool life would be expected. This trend is also found in the rest of the tests carried out. Considering the stresses it seems that chamfered-insert tools will endure better than round cutting edges in hard turning. This conclusion is opposite to the one obtained from temperature results.

Figure 5.12 shows the serrated chip obtained when machining at the cutting speed of 100 m min^{-1}, feed rate of 0.1 mm and cutting radius of 25 μm with AdvantEdge. To obtain this serrated chip with the Abaqus/Explicit model, smaller element dimensions should be employed (4–8 times smaller) [57]. This phenomenon appears in all of the calculations performed for the radius tool for

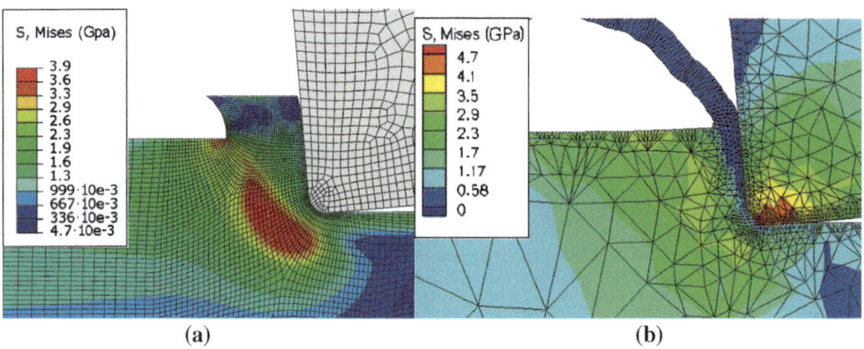

Figure 5.10 Mises equivalent stress fields in Abaqus (a) and AdvantEdge (b) at 100 m min^{-1} and 0.1 mm feed

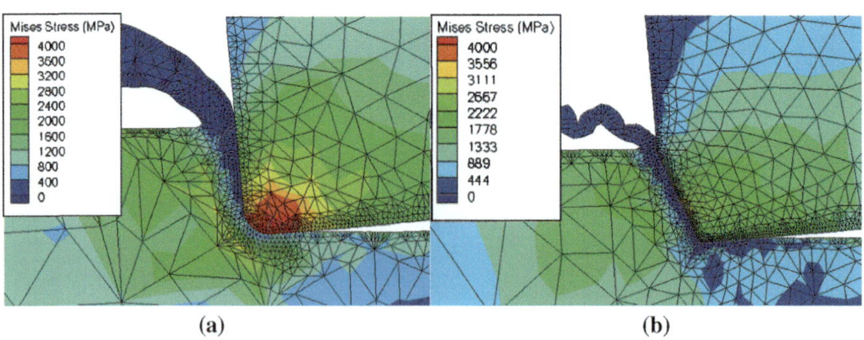

Figure 5.11 Mises equivalent stress plots of radius model (a) and chamfered model (b) in AdvantEdge at 100 m min^{-1} and 0.1 mm feed

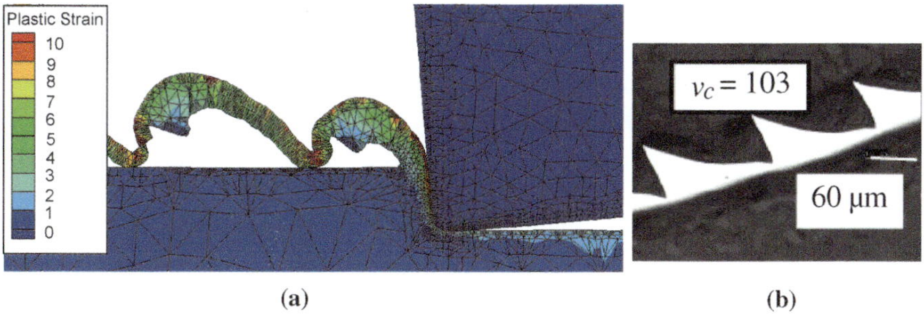

Figure 5.12 (a) Serrated chip obtained with the round cutting edge (AdvantEdge: cutting speed of 100 m min^{-1}, feed rate of 0.1 mm and cutting radius of 25 μm), and (b) serrated chip obtained at $v_c = 103$ m min^{-1} and feed rate of $f = 0.1$ mm

Figure 5.13 Cutting (X) and feed (Y) forces obtained

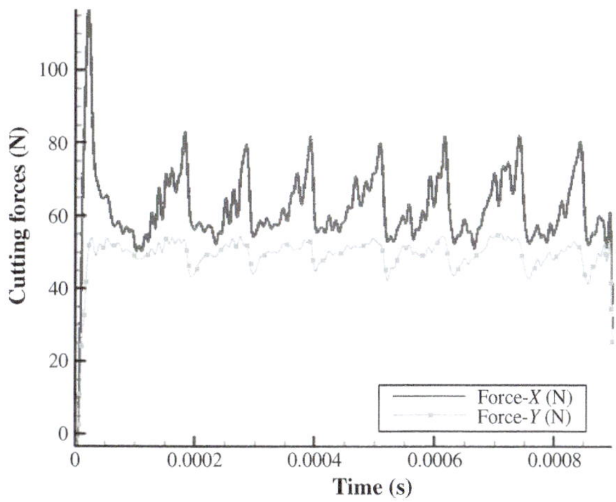

every feed and cutting speed and induces variations in the cutting forces obtained, since the amount of material machined varies continuously. In Figure 5.13 these variations can be observed: roughly 20 N in the cutting force and 10 N in the feed force.

The serrated chip is quite common in hard turning and has been observed in several research works [21].

5.3.3 Three-dimensional Finite-element Analysis of Hard Turning

In order to show residual stresses in machined surfaces, 3D simulations should be run. In this case, DEFORM-3D has been employed to provide residual stresses (as well as results for forces, temperatures, *etc.*) during hard turning. The experimental plan can be observed in Table 5.3. Four simulations were carried out with

DEFORM-3D, varying the cutting speed (v) from 120 to 140 m/min, and the feed rate (f) from 0.1 mm/rev to 0.15 mm. The depth of cut was kept constant at the value of 0.2 mm.

A comparison with previous research work with Abaqus/Explicit™ will be made: cutting speed (v) of 120 m/min., a feed rate (f) of 0.1 mm/rev and a depth of cut (p) of 0.2 mm [38]. Data about the Abaqus/Explicit model is shown in Table 5.3.

Table 5.3 Material properties and cutting condition for the process simulations (Abaqus/Explicit and DEFORM-3D)

			Abaqus/Explicit, DEFORM-3D
Material properties	Plasticity, Johnson–Cook law	A (MPa)	2482.4
		B (MPa)	1498.5
		n	0.19
		C	0.027
		m	0.66
	Inelastic heat fraction (β)		0.9
	Density (ρ) (kg m^{-3})	Workpiece (AISI 52100 62HRC)	7827
		Tool (PCBN)	3120
	Elasticity (E) (GPa)	Workpiece	201.33 f(T)
		Tool	680
	Conductivity (k) (W m^{-1} K^{-1})	Workpiece	43
		Tool	100
	Specific heat (c) (J kg^{-1} K^{-1})	Workpiece	458
		Tool	960
	Expansion (K^{-1})	Workpiece	11.5×10^{-6} f(T)
		Tool	4.9×10^{-6}
Contact	Thermal conductance (K_i) (W m^{-2} K^{-1})		1×10^8
	Heat partition coefficient (Γ)		0.5
	Friction coefficient (μ)		0.35 (Guo)
	Friction energy transformed into heat (η)		1
Process	Cutting speed (v) (m/min)		120/140
	Feed rate (f) (mm/rev)		0.1/0.15
	Depth of cut (p) (mm)		0.2
	Cutting-edge microgeometry		Chamfer 20° × 0.1 mm
	Nose radius (r_p) (mm)		0.8
	Rake angle (γ) (°)		–5
	Clearance angle (α) (°)		5
	Cutting-edge inclination angle (λ_s) (°)		–5
	Cutting-edge angle (K_s) (°)		0

Table 5.4 Experimental plan and results obtained for Abaqus/Explicit and DEFORM-3D

Test	Cutting forces			Tool	Workpiece			Chip		
	F_c (N)	F_f (N)	F_p (N)	Temp (K)	Temp (K)	ε	σ (GPa)	Temp (K)	ε	σ (GPa)
Vc_120_F_010 (Abaqus/Explicit)	154	58	138	1558	–	–	–	1483	–	–
Vc_120_F_010 (DEFORM-3D)	138	47	108	1065	963	4.8	1.39	1128	6.8	2.31
Vc_120_F_015 (DEFORM-3D)	192	59	147	1060	971	5.24	1.48	1139	6.3	2.19
Vc_140_F_010 (DEFORM-3D)	134	46	102	1156	993	5.72	1.63	1190	6.4	2.08
Vc_140_F_015 (DEFORM-3D)	188	56	142	1116	1012	6.3	1.7	1167	6.6	1.92

Table 5.5 shows the Von Mises stress contours obtained with the Abaqus model. The approach employed is as follows. Firstly, after an Abaqus/Explicit step is finished, the model is imported into Abaqus/Standard™. Once the import is completed, a new mesh is designed taking into account the deformed state of the previous mesh. For this task, a high-performance finite-element pre-processor is employed, Altair Hypermesh 8.0™, which permits the creation of surfaces from a previous mesh. After the desired mesh is created, a mapping calculation is made in Abaqus/Standard, in order to interpolate the solution onto the new mesh from the output databases generated with the old mesh. Finally, the analysis is imported into Abaqus/Explicit and a new step is performed. This procedure is repeated until desired steady-state solution has been reached [38].

The ALE formulation employed for the final step (see Table 5.5) makes the material enter the workpiece mesh and exit it via different Eulerian boundary surfaces.

Table 5.5 Initial mesh, remeshing steps, and final steady-state step for the ALE finite-element model for 3D hard turning

Step	Total time	
4 Fully Eulerian boundaries	0.165 ms (0.33 mm)	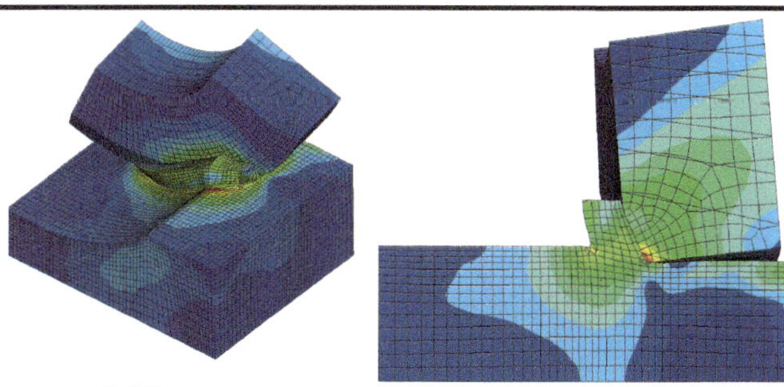

Thanks to this new hybrid formulation employed, artificial criteria (physical or geometrical) to generate the chip or an initial chip design are not needed. Only a short initial remeshing stage is needed to obtain the final chip shape. Such features, probably the major advantages of the presented model over the ones reported in the literature [18, 59], avoid the introduction of arbitrary effects on the obtained results, making the simulation more robust. This point gains special relevance when trying to forecast the state of the machined surface and residual stresses.

More detailed information about this work can be found in [38, 58].

Both tool and workpiece are considered as deformable, in order to analyze the stresses that appear in the tool. Only elastic properties are used for the tool.

Due to the limitations imposed by the importing options in Abaqus/Standard, in these 3D simulations the Johnson–Cook law has been discretized into tabular data. The number of points employed to describe the material behavior is 15,525, with 25 levels for the strain, 23 for the strain rate, and 27 for the temperature.

The Coulomb friction law has been employed for the modeling of the tool–chip interface contact. Heat transfer is allowed on the tool–chip contact area.

The elements employed are C3D8RT, eight-node bricks with tri-linear displacement, temperature calculation, and hourglass control. Their size varies from 2 μm to 30 μm depending on the zone. Table 5.3 shows the input parameters for the hard-turning simulation [59].

5.3.4 Three-dimensional Finite-element Analysis of Hard Turning: Results and Discussion

The simulation was run until the steady state was reached: 0.133 ms in the case of Abaqus/Explicit, and 0.25 ms in the case of DEFORM-3D. Results for forces and temperatures in Figures 5.14 and 5.15, respectively, confirm that the steady state has been reached. As seen in Figure 5.14, for DEFORM-3D the cutting force seems to stabilize around 140 N with small perturbations of about 2 N.

The cutting force obtained from the hard-turning simulation with Abaqus/Explicit is 154 N, while the feed force has a value of 58 N. The radial force has a value of 138 N, which is almost as big as the cutting force. For DEFORM-3D simulations, values of cutting force vary from 192 to 134 N, while the radial force varies from 147 to 108 N. The feed force varies between the values of 47 N and 59 N. It can be observed (Figure 5.16), that as the feed increases, the values of the forces raise in a similar way. The results for the cutting force match qualitatively well with results obtained experimentally by Huang and Liang [58]. However, differences are more significant when comparing feed force and passive (radial) forces. Lower values are obtained in FEM. It is clear that the use of better friction models with proper identification friction parameters would have given results closer [58] to the experimental ones.

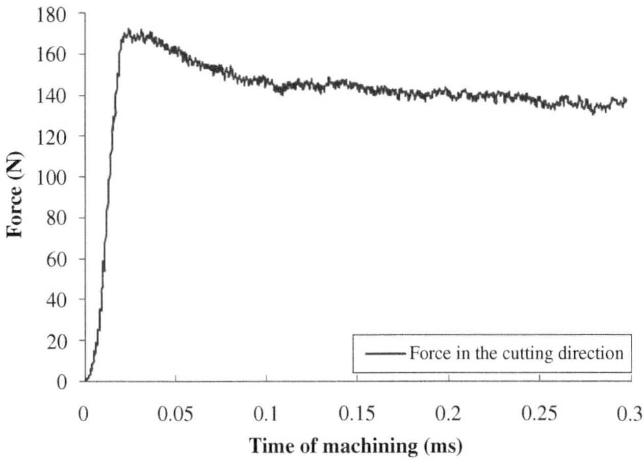

Figure 5.14 Cutting forces for DEFORM-3D

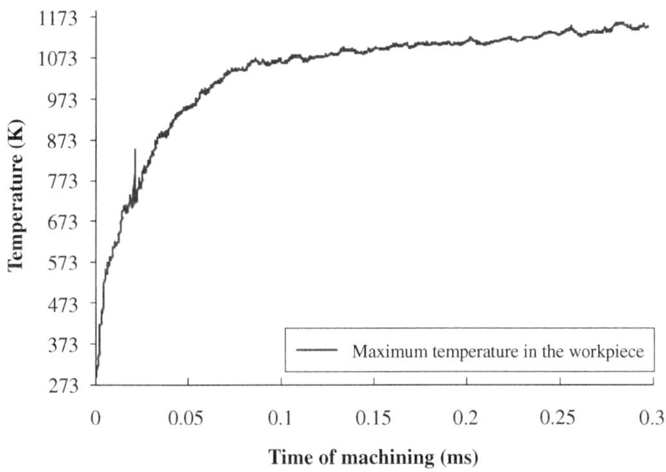

Figure 5.15 Workpiece maximum temperature plot for DEFORM-3D

What is more, looking at Figure 5.17, it can be said that the rise in the cutting speed causes a slight decrease in the values of the cutting forces. This means that the increase in the forces depends more on the feed, rather than on the cutting speed.

Figure 5.18 to 5.22 show the temperature fields for the hard-turning simulation in the tool and workpiece for the five cases studied. In the case of the DEFORM results, tool temperature varies form 1065 K to 1116 K. Differences with Abaqus/Explicit are again quite relevant (>400 K) and are due again, as in 2D, to differences in input parameters, solver, *etc.*

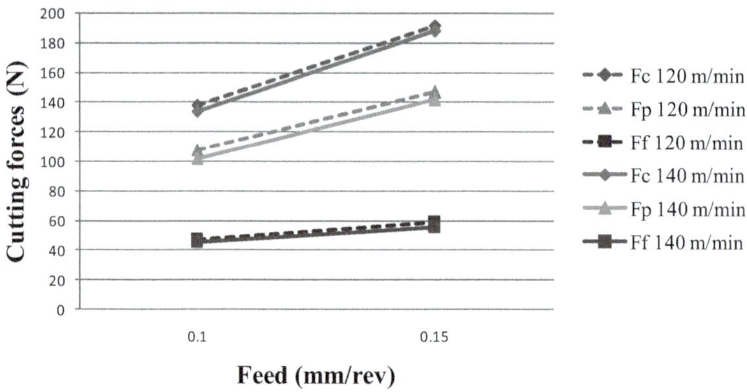

Figure 5.16 Influence of feed on cutting forces for DEFORM-3D simulations

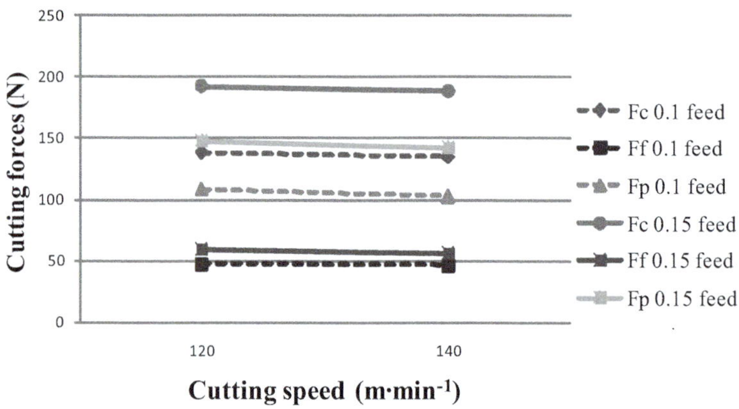

Figure 5.17 Influence of cutting speed on cutting forces for two different feeds for DEFORM-3D simulations

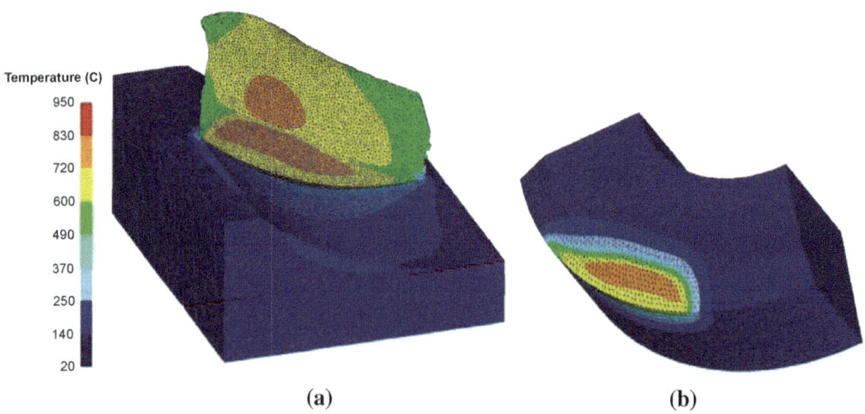

Figure 5.18 Temperature (°C) contour maps in the workpiece (a) and tool (b) for Vc_120_F_010 (DEFORM-3D) simulation

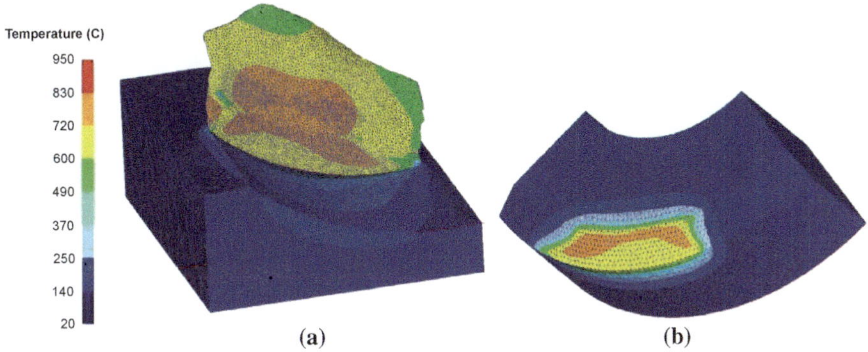

Figure 5.19 Temperature (°C) contour maps in the workpiece (a) and tool (b) for Vc_120_F_015 (DEFORM-3D) simulation

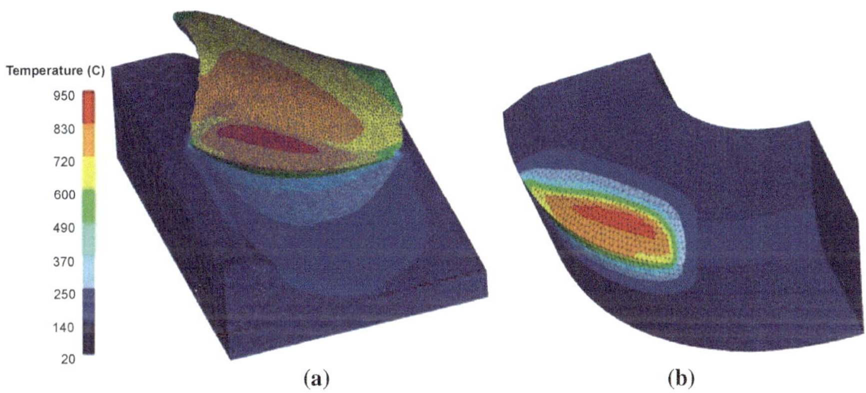

Figure 5.20 Temperature (°C) contour maps in the workpiece (a) and tool (b) for Vc_140_F_010 (DEFORM-3D) simulation

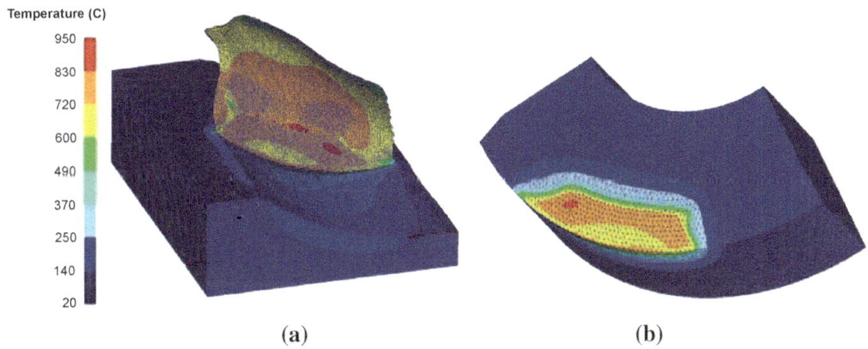

Figure 5.21 Temperature (°C) contour maps in the workpiece (a) and tool (b) for Vc_140_F_015 (DEFORM-3D) simulation

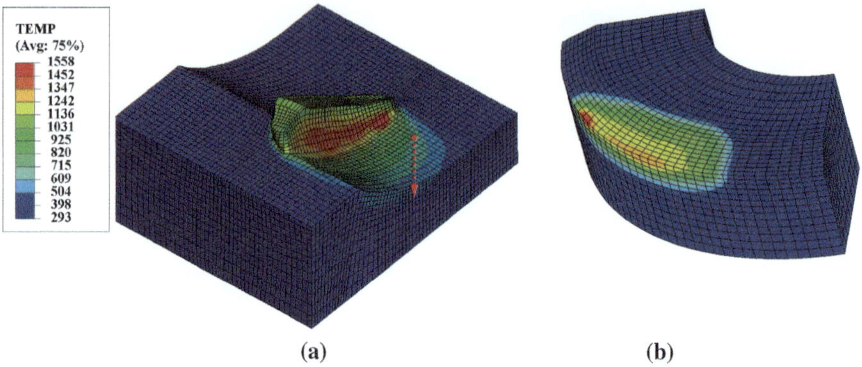

Figure 5.22 Temperature (K) contour maps in the workpiece (a) and tool (b) for Vc_120_F_010 (Abaqus/Explicit) simulation; *arrow* indicates the path used for the residual-stress extraction

Figure 5.23 shows the temperature results obtained depending on the feed rate. It is observed that when feed rate rises, the values of temperature in the tool seem to decrease slightly, while workpiece temperatures appear to increase.

On the other hand, when the cutting speed is increased, there is a significant increase in temperature results, as can be observed in Figure 5.24.

One of the main advantages of 3D simulations is that it is possible to analyze the variable state of the chip formation process along the cutting edge. Figures 5.25 and 5.26 show the plastic strain and Von Mises equivalent stress contour maps for different cutting planes (perpendicular to the cutting edge) for the test with a cutting speed of $v_c = 120\,\mathrm{m\ min^{-1}}$. The maximum plastic strain reaches the value of 6, and is generated where the undeformed chip thickness is a maximum.

On the other hand, the maximum stress appears in the tool, in the central zone of the chip width.

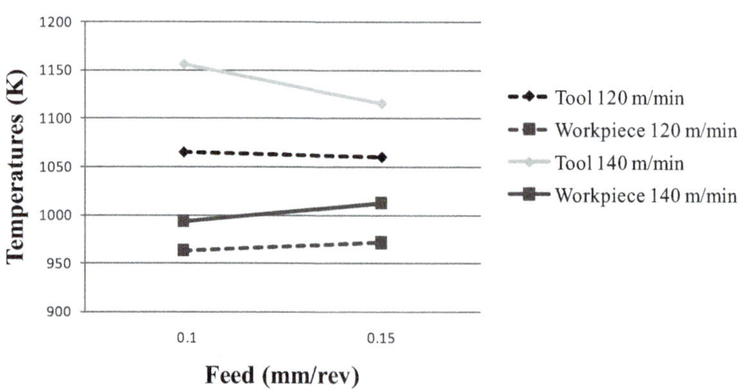

Figure 5.23 Influence of feed on temperatures for DEFORM-3D simulations

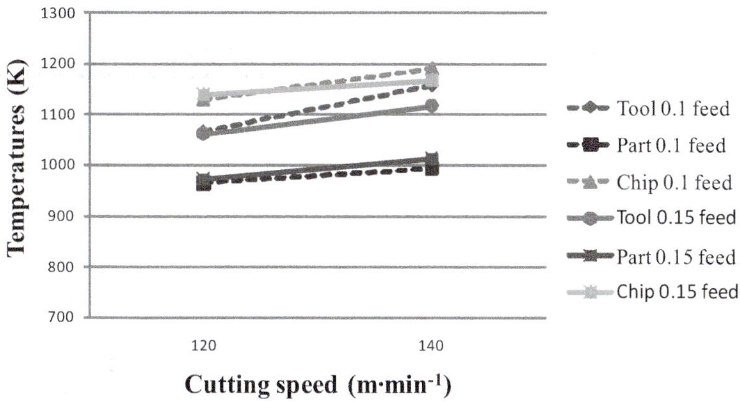

Figure 5.24 Influence of cutting speed on temperatures for two different feeds for DEFORM-3D simulations

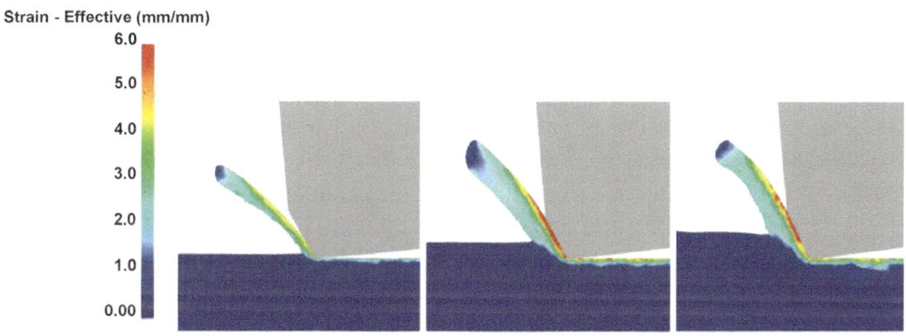

Figure 5.25 Plastic-strain equivalent stress for three different cross-sections along the cutting edge for Vc_120_F_010 (DEFORM-3D)

Figure 5.26 Von Mises equivalent stress for three different cross-sections along the cutting edge for Vc_120_F_010 (DEFORM-3D)

In DEFORM-3D simulations, the tool is rigid so it is impossible to analyze the contact pressure in the tool. However, Figures 5.27–5.31 show the contact pressure fields that appear in the workpiece. Comparison of Figure 5.27 and 5.29 (feed of 0.10 mm/rev) and Figure 5.28 and 5.30 (feed of 0.15 mm/rev) reveals that feed rate does not have a significant influence on contact pressure; similar values are obtained in all the cases (close to 8–9 GPa). In Figure 5.31 contact pressure obtained with the Abaqus model is shown. Values close to 8–9 GPa are obtained here as well.

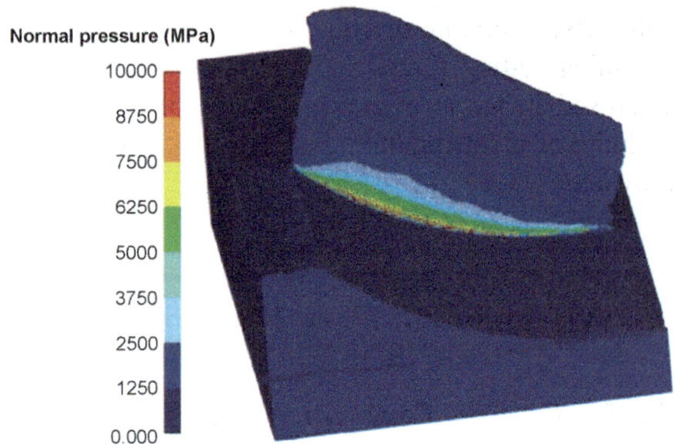

Figure 5.27 Contact pressure on the workpiece for Vc_120_F_010 (DEFORM-3D)

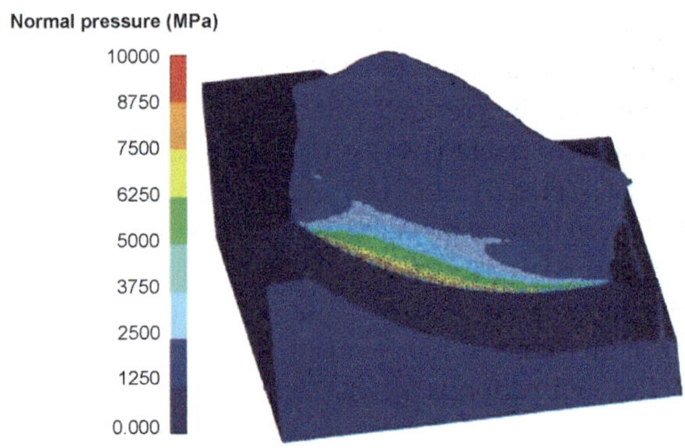

Figure 5.28 Contact pressure on the workpiece for Vc_120_F_015 (DEFORM-3D)

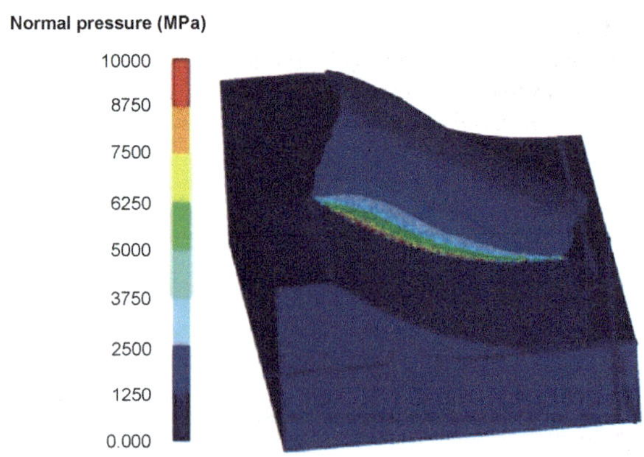

Figure 5.29 Contact pressure on the workpiece for Vc_140_F_010 (DEFORM-3D)

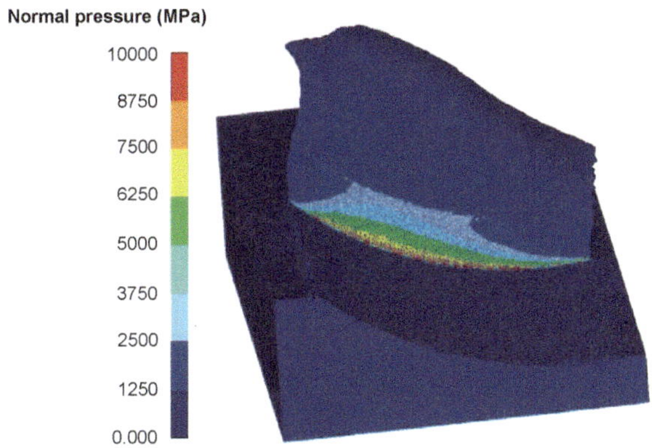

Figure 5.30 Contact pressure on the workpiece for Vc_140_F_015 (DEFORM-3D)

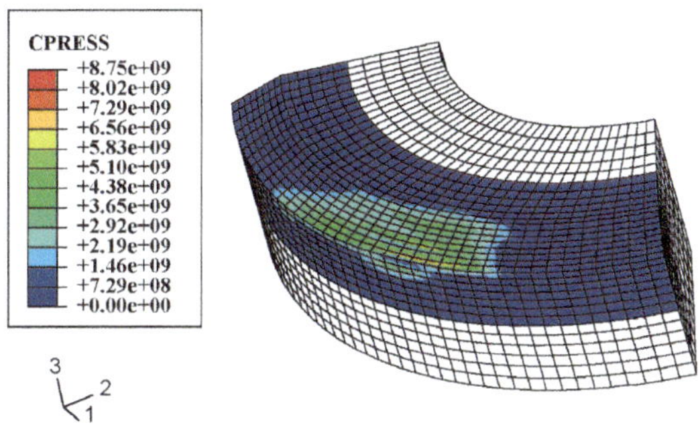

Figure 5.31 Contact pressure on the workpiece for Vc_120_F_010 (Abaqus/Explicit)

As explained earlier, an important advantage of the Abaqus model is that it does not delete any element from the model, so the superficial state of the workpiece can be estimated.

Figure 5.33 shows the residual-stress components corresponding to the ones which would be measured in a hole-drilling test. That is, only the planar stress state corresponding to the machined surface will be analyzed. For this task, stress values from a path (see Figure 5.22) have been extracted, corresponding to the material that would not be removed in the next revolution. The component S11 would correspond to the stress in the feed direction, component S33 to the cutting direction, and component S13 to the shear stress in the plane (see Figure 5.32).

Their values vary from –400 Mpa for S13 to –1200 Mpa for S33. That is, all of them are negative (compressive) in the machined surface, which matches roughly with experimental data found in Liu *et al.* [60]. As can be observed in Figure 5.34, according to experimental measures found in Liu *et al.* [60], a slightly compressive stress can be found at the machined surface, while the compressive stress becomes higher with the depth, until 15 μm is reached. However, this is not the case when analyzing the results of other authors such as Dahlman *et al.* [71], where tractive stresses are obtained at the machined surface (Figure 5.35).

Figure 5.32 Residual stresses in facing [60]

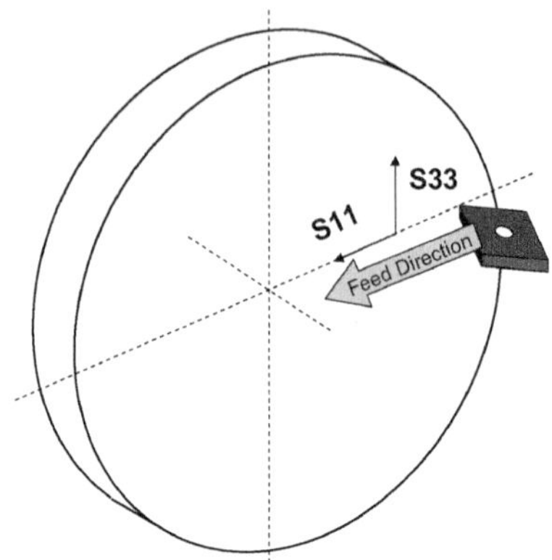

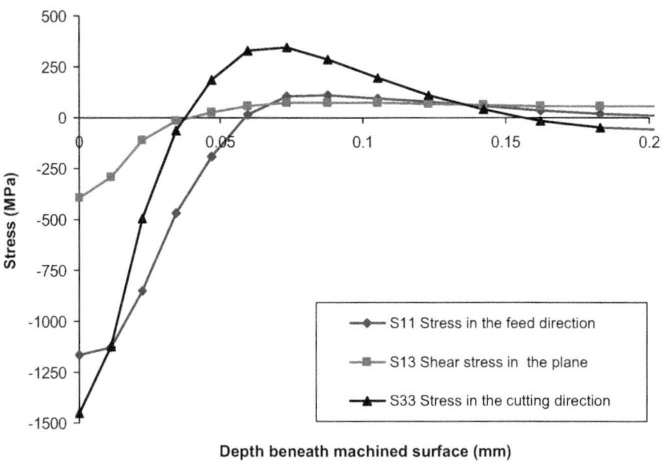

Figure 5.33 Stress components along the depth of the workpiece for Vc_120_F_010 (Abaqus/Explicit)

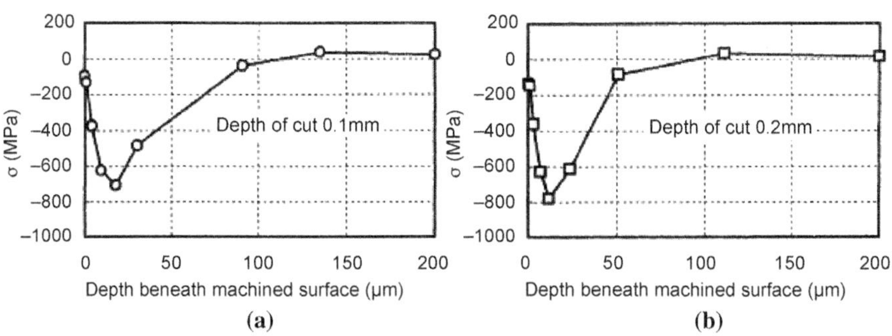

Figure 5.34 Residual stress found by Liu *et al.* [60] in the cutting direction for the same cutting conditions ($v = 120\,\mathrm{m\,min^{-1}}$; $f = 0.10\,\mathrm{mm}$). Depths of cut: (a) 0.1 mm, and (b) 0.2 mm

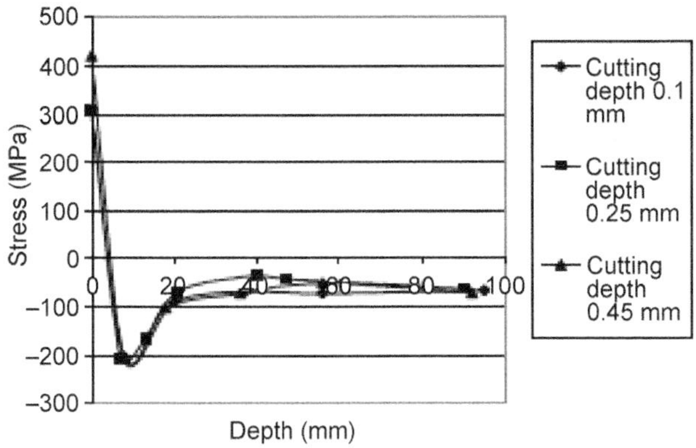

Figure 5.35 Residual stress found by Dahlman *et al.* [71] in the cutting direction for the same cutting conditions ($v = 110\,\text{m min}^{-1}$, $f = 0.15$ mm) for different depths of cut (0.1 mm, 0.25 mm, and 0.45 mm)

On the other hand, the compressive residual stresses found in the Abaqus/Explicit simulation are a maximum in the machined surface. This could be due to the high temperature found on this surface (570 K) which could generate compressive stresses due to the thermal expansion. This fact suggests that a cooling and stress-relaxation step should be carried out in order to obtain more realistic residual stresses.

To analyze residual-stress results with DEFORM, new simulations have been launched where cooling effects are considered. When the workpiece reaches a temperature of 293 K (20 °C), stress values have been obtained along the depth of the workpiece. In DEFORM-3D, the output variable Stress-X corresponds to stress in the feed direction, while Stress-Y corresponds to stress in the cutting direction and Tau-XY to shear stress in the plane. Due to the reference system employed, the values of Stress-Y have to be the opposite ones. Thus, all the values of residual stresses on the machined surface are tractive. As can be seen in Figures 5.36–5.39 their maximum values vary from 800 MPa for stress in the cutting direction to 600 MPa for shear stress in the plane.

Looking at the results given by the different residual-stress tests, the following conclusions can be drawn:

- If the feed is increased, the slope of the curves is smaller. This effect can be seen in Figures 5.37 and 5.39, the tests where the feed is 0.15 mm/rev. In contrast, in those tests where the feed is 0.10 mm/rev (Figures 5.36 and 5.38), the slope of the curves is higher and the stress tends to negative values quicker.
- On the other hand, the rise in the cutting speed and feed rate does not seem to change significantly the obtained results.

Considering the experimental results for the residual stresses found by Dahlman *et al.* [71], it can be said that DEFORM-3D results match qualitatively well. In these experimental results, stresses are tractive at the machined surface and differ from the results of Liu *et al.* [60]. At the depth of 25 μm the stresses become compressive.

Taking into consideration the quantitivive results, a tractive stress of 300 MPa has been obtained on the machined surface by Dahlman *et al.* [71], while values close to 700–750 MPa are obtained in simulations. This data does not match the simulation data, but it does for the compressive stress of 200 MPa at the minimum point.

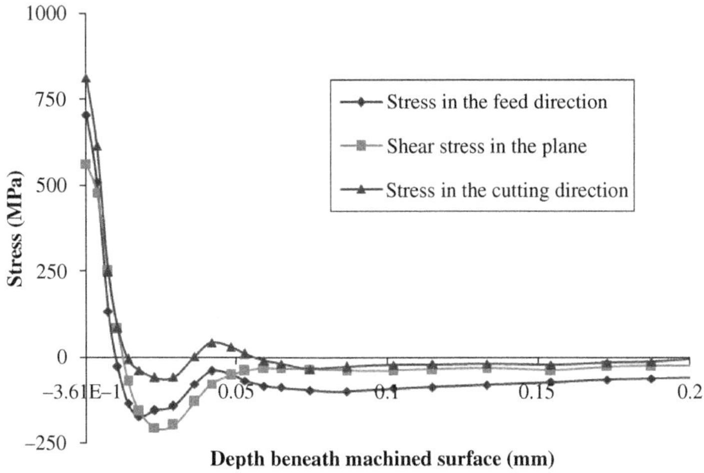

Figure 5.36 Stress components along the depth of the workpiece after cooling step for Vc_120_F_010 (DEFORM-3D)

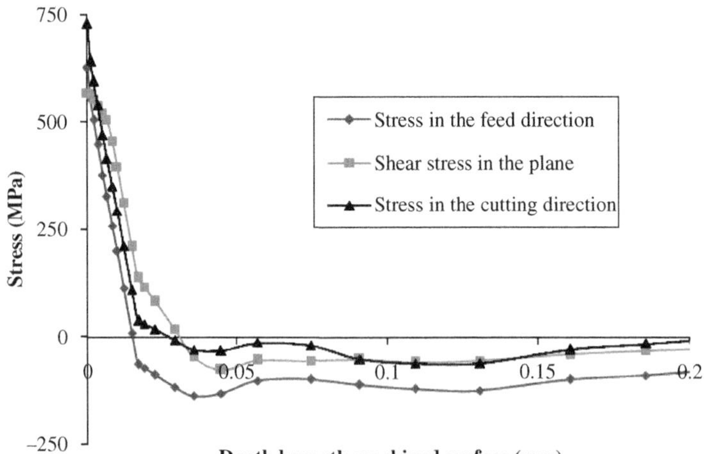

Figure 5.37 Stress components along the depth of the workpiece after cooling step for Vc_120_F_015 (DEFORM-3D)

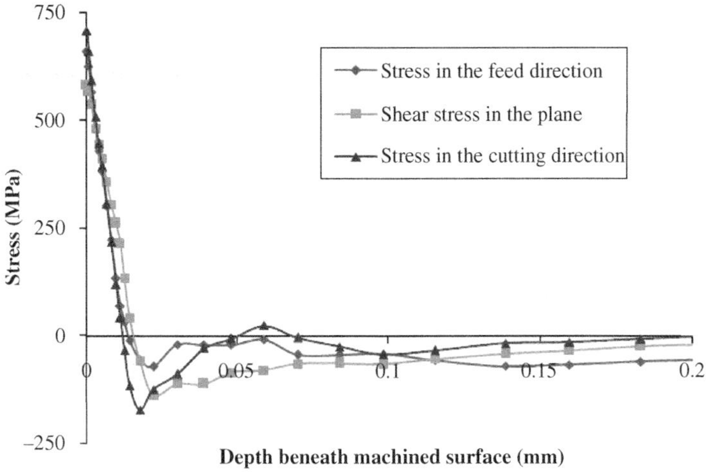

Figure 5.38 Stress components along the depth of the workpiece after cooling step for Vc_140_F_010 (DEFORM-3D)

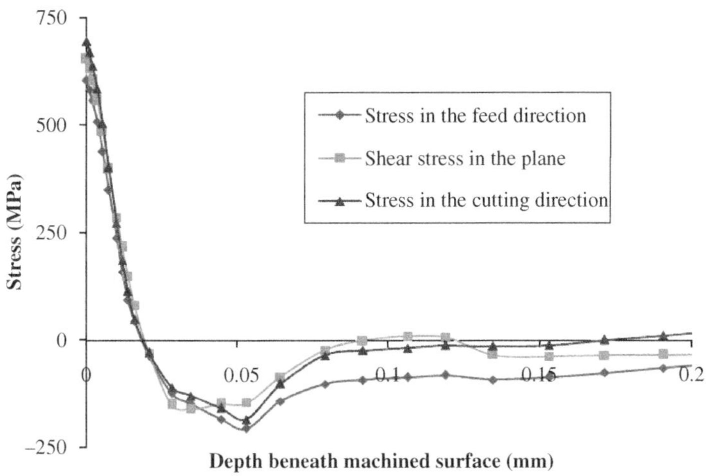

Figure 5.39 Stress components along the depth of the workpiece after cooling step for Vc_140_F_015 (DEFORM-3D)

5.4 Conclusions

The following conclusions can be drawn:

- FEM can give quite interesting qualitative values about the influence of input parameters on results like temperature, stresses, pressure, *etc.* that are fairly difficult to be measured experimentally (in particular in 3D).
- 3D modeling would be needed to meet industrial requirements regarding residual stresses, tool wear, *etc.*

The uncertainty in identification of entry parameters can have a significant influence on FEM uncertainty.

- In 2D analysis:
  - Remarkable differences are observed in quantitative values between Abaqus and AdvantEdge. This was expected due to the difficulties of employing the same input parameters in both models.
  - Round cutting edges seem to give lower values of tool temperature but higher values of stresses.
  - In AdvantEdge a serrated-chip-like phenomenon has been observed. However, in Abaqus this phenomenon has not been observed due to the large element size.
- In 3D analysis:
  - Quite similar results for forces have been obtained in DEFORM and Abaqus. It can be said that cutting forces match quite well with experimental results, while the feed and passive (radial) force differ more from experimental results. Proper friction coefficients and models should be employed if more precise quantitative values are expected to be achieved. In the case of temperatures, more significant differences have been obtained between Abaqus and DEFORM (close to 300 K);
  - There are remarkable differences between DEFORM and Abaqus for residual-stress values. As stated in the previous section, it seems that DEFORM-3D matches better than Abaqus/Explicit model with the experimental curves obtained by other researchers.

Acknowledgments The authors would like to thank the Basque and Spanish Governments for the financial support given to the projects: Manufacturing 0,0 (code IE08–2222), Manufacturing 0,0 II (code IE09–254) and Metincox (DPI2009–14286-C02–0 and PI2010-11).

References

[1] Koenig W, Berktold A, Koch K-F (1993) Turning versus grinding – a comparison of surface integrity aspects and attainable accuracies. Ann CIRP 42(1):39–43

[2] Balart MJ, Bouzina A, Edwards L, Fitzpatrick ME (2004) The onset of tensile residual stresses in grinding of hardened steels. Mater Sci Eng 367(1–2):132–142

[3] Guo YB, Barkey-Mark E (2004) Modeling of rolling contact fatigue for hardened machined components with process-induced residual stress. Int J Fatigue 26(6):605–613

[4] Guo YB, Sahni J (2004) A comparative study of hard turned and cylindrically ground white layers. Int J Mach Tools Manuf 44(2/3):135–145

[5] Tönshoff HK, Karpuschewski B, Borbe C (1998) Hard machining – state of research. In: Proceedings international CIRP/VDI-conference on high performance tools, Düsseldorf (Germany), pp 253–277

[6] Chou YK, Song H (2004) Tool nose radius effects on finish hard turning. J Mater Process Technol 148(2):259–268

[7] Sukaylo VA, Kaldos A (2004) Development and verification of a computer model for thermal distortions in hard turning. J Mater Process Technol 155–156:1821–1827
[8] Umbrello D, Jayal AD, Caruso S, Dillon OW, Jawahir IS (2009) Modeling of white and dark layers formation in orthogonal machining of hardened AISI 52100 steel. Proceedings of the 12th conference on modelling machining operations, Donostia-San Sebastián (Spain), pp 655–662
[9] Strenkowski JS, Moon K (1990) Finite element prediction of chip geometry and tool/workpiece temperature distributions in orthogonal metal cutting. J Eng Ind 112:313–318
[10] Denkena B, Jung M, Müller C, Kramer N (2004) Characterisation of white layers inflicted by mechanical and thermal loads within manufacturing processes. In: 7th international symposium on advances in abrasive technology, Bursa (Turkey)
[11] Rech J, Moisan A (2003) Surface integrity in finish hard turning of case-hardened steels. Int J Manuf 43(5):543–550
[12] Barry J, Byrne G (2002) TEM study on the surface white layer in two turned hardened steels. Mater Sci Eng A 325:356–364
[13] Warren A-W, Guo Y-B, Weaver M-L (2006) The influence of machining induced residual stress and phase transformation on the measurement of subsurface mechanical behavior using nanoindentation. Surf Coat Technol 200:3459–3467
[14] Guo Y-B, Sahni J (2004) A comparative study of hard turned and cylindrically ground white layers. Int J Mach Tools Manuf 44:135–145
[15] Ekinovic S, Doninsek S, Jawahir I-S (2004) Some observations of the chip formation process and the white layer formation in high speed milling of hardened steel. Mach Sci Technol 8(2):327–340
[16] Al-Wardany T-I, Kishawy H-A, Elbestawi M-A (2000) Surface integrity of die material in high speed hard machining. Part 1: Micrographical analysis. J Manuf Sci Eng 122:620–631
[17] Smith S, Melkote SN, Lara-Curzio E, Watkins TR, Allard L, Riester L (2007) Effect of surface integrity of hard turned AISI 52100 steel on fatigue performance. Mater Sci Eng A 459: 337–346
[18] Davies MA, Chou YK, Evans CJ (1996) On chip morphology, tool wear and cutting mechanics in finish hard turning. Ann CIRP 45(1):77–82
[19] Özel T, Karpat Y, Srivastava A (2008) Hard turning with variable micro-geometry PcBN tools. Ann CIRP 57(1):73–76
[20] Karpat Y, Özel T (2007) 3D FEA of hard turning: investigation of PcBN cutting tool micro-geometry effects. Trans NAMRI/SME 35:9–16
[21] Karpat Y, Özel T, Sockman J, Shaffer W (2007) Design and analysis of variable micro-geometry tooling for machining using 3D process simulations. In: Proceedings of international conference on smart machining systems, March 13–15. National Institute of Standards and Technology, Gaithersburg, MD
[22] Poulachon, G, Moisan, AL, Jawahir, IS (2007) Evaluation of chip morphology in hard turning using constitutive models and material property data. J Manuf Sci Eng 129:41–47
[23] Klocke F, Brinksmeier E, Weinert K (2005) Capability profile of hard cutting and grinding processes. Ann CIRP 54(2):557–580
[24] Klocke F, Kratz H (2005) . Ann CIRP 54(1):47–50
[25] Attanasio A, Ceretti E, Rizzuti S, Umbrello D, Micari F (2008) 3D finite element analysis of tool wear in machining. Ann CIRP 57(1):61–64
[26] Okushima K, Kakino Y (1971) Residual stress produced by metal cutting. Ann CIRP 20(1):13–14
[27] Tay AO, Stevenson MG, Davis GDV, Oxley PLB (1974) Using the finite element method to determine temperature distributions in orthogonal machining. Proc Inst Mech Eng 188:627–638
[28] Tay AO, Stevenson MG, Davis GDV, Oxley PLB (1976) A numerical method for calculating temperature distributions in machining, from force and shear angle measurements. Int J Mach Tool Des Res 16:335–349

[29] Ng E, Aspinwall DK (2000) Hard part machining AISI H13 (≈50HRc) using AMBORITE AMB90: A finite element modeling approach. Ind Diam Rev 4:305–310
[30] Merchant E (1944) Basic mechanics of the metal-cutting process. Transaction of the ASME J App Mech 66:168–175
[31] Lee EH, Shaffer BW (1951) The theory of plasticity applied to a problem of machining. ASME J App Mech 73:404–413
[32] Armarengo EJA, Brown RH (1969) The machining of metals. Prentice-Hall, Englewood Cliffs, NJ
[33] Dudzinski D, Molinari A (1997) A modelling of cutting for viscoplastic materials. Int J Mech Sci 39(4):369–389
[34] Altintas Y (2000) Manufacturing Automation. Metal cutting mechanics, machine tool vibrations and CNC design. Cambridge University Press
[35] Leopold J (1998) FEM modeling and simulation of 3D chip formation. In: Proceedings of the 1st CIRP international workshop on modeling of machining operations, May, Atlanta, GA, pp 235–245
[36] Leopold J, Semmler U, Hoyer K (1999) Applicability, robustness and stability of the finite element analysis in metal cutting operations. In: Proceedings of the 2nd CIRP international workshop on modeling of machining operations, Nantes (France), pp 81–94
[37] Madhavan V, Chandrasekar S, Farris TN (2000) Machining as a wedge indentation. J Appl Mech 67:128–139
[38] Arrazola PJ, Özel T (2009) Finite element modelling of machining processes. In: Özel T, Davim JP (eds) Intelligent machining: modeling and optimization of the machining processes and systems. ISTE, London, pp 125–163
[39] Arrazola P.J, Garay A, Villar A, San Juan M, Santos FJ, Martín O (2009) Modelización del proceso de formación de viruta con elementos finitos: Identificación de la ley de comportamiento material pieza. Proc of the 3rd CISIF-MESIC. Alcoy
[40] Strenkowski JS, Carroll JT (1985) A finite element model of orthogonal metal cutting. J Eng Ind 107:349–354
[41] Ng E, Aspinwall DK, Brazil D, Monoghan J (1999) Modelling of temperature and forces when orthogonally machining hardened steel. Int J Mach Tools Manuf 39:885–903
[42] Ueda K, Manabe K (1993) Rigid-plastic FEM analysis of three-dimensional deformation field in chip formation process. Ann CIRP 42(1)35–38
[43] Shirakashi T (2003) FEM simulation analysis on ductile mode glass machining process. In: Proceedings of the 6th international ESAFORM conference on material forming, Salerno (Italy), pp 539–542
[44] Söhner J, Weule H, Biessinger F, Schulze V, Vöhringer (2001) Examinations and 3D-simulations of HSC face milling process. In: Proceedings of the 4th CIRP international workshop modelling of machining operations, Delft, The Netherlands, August, pp 111–116
[45] Marusich TD, Ortiz M (1995) Modeling and simulation of high-speed machining. Int J Num Meth Eng 38:3675–3694
[46] Marusich TD, Brand CJ, Thiele JD (2002) A methodology for simulation of chip breakage in turning processes using an orthogonal finite element model. In: Proceedings of the 5th CIRP international workshop on modeling of machining operations, May, West Lafayette (USA), pp 139–148
[47] Fourment L, Bouchard PO, Bay F, Chenot JL (1998) Numerical simulation of chip formation and crack propagation during non-steady cutting process. In: Proceedings of the 2nd CIRP international workshop on modelling of machining operations, Nantes (France), pp 108–123
[48] Ceretti E, Lazzaroni C, Menegardo L, Altan T (2000) Turning simulations using a three-dimentional FEM code. J Mater Process Technol 98:99–103
[49] Ceretti E (1999) Numerical study of segmented chip formation in orthogonal cutting. In: Proceedings of the 2nd CIRP international workshop on modelling of machining operations, Nantes (France), pp 124–132

[50] Carroll GJT, Strenkowski JS (1988) Finite element models of orthogonal cutting with application to single point diamond turning. Int J Mech Sci 30(12):pp 899–920
[51] Strenkowski JS, Althavale SM (1997) A partially constrained eulerian orthogonal cutting model for chip control tools. J Manuf Sci Eng, 119:681–688
[52] Maekawa K, Kubo A, Childs THC (2001) A friction model for freemachining steels and its applicability to machinability analysis. Key Eng Mater 196:79–90
[53] Pantale O, Rakotomalala R, Touratier M, Hakem N (1996) A three dimensional Numerical Model of orthogonal and oblique metal cutting processes. Eng Syst Des Anal ASME-PD 75:199–205
[54] Movahhedy MR, Gadala MS, Altintas, Y (2000) FE modeling of chip formation in orthogonal metal cutting process: An ALE approach. Mach Sci Technol 4:15–47
[55] Arrazola PJ, Özel T (2010) Investigations on the effects of friction modeling in finite element simulation of machining. Int J Mech Sci 52(1):31–42
[56] Llanos I, Villar JA, Urresti I, Arrazola PJ (2009) Finite element modeling of oblique machining using an arbitrary Lagrangian-Eulerian formulation. Mach Sci Technol 13:122. doi: 10.1080/10910340903237921
[57] Arrazola PJ, Villar A, Ugarte D, Marya S (2007) Serrated chip prediction in finite element modeling of the chip formation process. Mach Sci Technol 11(3):367–390
[58] Arrazola PJ, Ugarte D, Domínguez X (2008) A new approach for the friction identification during machining through the use of finite element modelling. Int J Mach Tools Manuf 48:73–183
[59] Guo YB, Liu CR (2002) 3D FEA modeling of hard turning. ASME J Manuf Sci Eng 124:189–199
[60] Yen Y-C, Jain A, Altan T (2004) A finite element analysis of orthogonal machining using different tool edge geometries. J Mater Process Technol 146(1):72–81
[61] Chen L, El-Wardany TI, Nasr M, Elbestawi, MA (2006) Effects of edge preparation and feed when hard turning a hot work die steel with polycrystalline cubic boron nitride tools. Ann CIRP 55(1):89–92
[62] Aurich JC, Bil H (2006) 3D finite element modelling of segmented chip formation. Ann CIRP 55(1):47–50
[63] Kountanya R, Al-Zkeri I, Altan T (2009). Effect of tool edge geometry and cutting conditions on experimental and simulated chip morphology in orthogonal hard turning of 100Cr6 steel. J Mater Process Technol 209:5068–5076
[64] Umbrello D, Filice L (2009) Improving surface integrity in orthogonal machining of hardened AISI 52100 steel by modeling white and dark layers formation. CIRP Ann Manuf Technol 58(1):73–76
[65] Umbrello D, Rizzuti S, Outeiro J.C, Shivpur, R, M'Saoubi R (2008). Hardness-based flow stress for numerical simulation of hard machining AISI H13 tool steel. J Mater Process Technol 199:64–73
[66] Piendl S, Aurich JC, Steinicke M (2005) 3D finite-element simulation of chip formation in turning. In: 5th CIRP international workshop on modeling of machining operations, pp 225–233
[67] Warnecke G, Oh J-D (2002) A new thermo-viscoplastic material model for finite-element-analysis of the chip formation process. Ann CIRP 51(1):79–82
[68] Johnson GR, Cook WH (1983) A constitutive model and data for metals subjected to large strains, high strain rates and high temperatures. In: Proceedings of the 7th international symposium on ballistics, The Hague (The Netherlands), pp 541–547
[69] Huang, Y, Liang, SY (2005) Modeling of cutting forces under hard turning conditions considering tool wear effect. Trans ASME 127:262–270
[70] Liu M, Takagi J, Tsukuda A (2004) Effect of tool nose radius and tool wear on residual stress distribution in hard turning of bearing steel. J Mater Process Technol 150:234–241
[71] Dahlman P, Gunnberg F, Jacobson M (2003) The influence of rake angle, cutting feed and cutting depth on residual stresses in hard turning. Chalmers University of Technology, Gothenburg (Sweden)

Chapter 6
Computational Methods and Optimization

R. Quiza and J.P. Davim

This chapter aims to illustrate the application of computer-based techniques and tools in modelling and optimization of hard-machining processes. An overview of the current state-of-the-art in this wide topic is reflected. Computational methods are explained not only for modelling the relationships between the variables in the cutting process, but also for optimizing the most important parameters. The characteristics of these techniques are exposed and their advantages and shortcomings are compared. Foreseen future trends in this field are presented.

6.1 Introduction

Mathematical modelling of cutting processes is very important, not only for understanding the nature of the process itself, but also for planning and optimizing the machining operations. Nevertheless, hard machining involves many complex and nonlinear relationships between different variables and parameters. Modelling of these relationships is a difficult task.

Although the analytical models help to provide better insight into the underlying physical nature of the cutting process in hard machining, they are usually less satisfactory in modelling variables due to simplifications and assumptions [1]. Therefore, empirical models must be used instead; however, identification of useful rela-

R. Quiza
Department of Mechanical Engineering, University of Matanzas, Autopista a Varadero, km 3½, Matanzas 44740, Cuba
e-mail: quiza@umcc.cu

J.P. Davim
Department of Mechanical Engineering, University of Aveiro, Campus Santiago, 3810-193 Aveiro, Portugal
e-mail: pdavim@ua.pt

tionships from raw experimental data is not easy. Historically, statistical tools (such as DoE, sampling and multiple regression) have been widely used, but application of these techniques to hard machining is far from a satisfactory success.

In recent years, artificial intelligent tools have gained popularity in the research community, as shown by the increasing number of publications on these topics. The so-called soft computing techniques (*i.e.*, artificial neural networks, fuzzy logic, neuro-fuzzy systems, *etc.*) are the most used approaches in hard-machining modelling.

Additionally, optimization of cutting parameters, although quite important for planning efficient machining processes, is a complicated target, challenged not only by the complex nature of the involved phenomena but also by the need of carefully defining realistic optimization objectives, and developing and implementing powerful and versatile optimization techniques. In this sense, stochastic optimization techniques, mainly evolutionary algorithms, have been widely reported in the recent literature.

This chapter intends to present a panoramic view of the current application of computational tools in hard-machining modelling and optimization. With this objective, it is divided in two sections. The first one exposes the computational techniques for modelling, including not only intelligent techniques but also other more conventional approaches that have proved to be effective for this purpose. The second one describes the hard-machining optimization problem and reviews the recently used tools, comparing their performance. A case study is included in order to illustrate the combination of neural networks and genetic algorithms (GAs) in solving a turning optimization problem. Finally, the future trends in these fields are roughly foreseen.

6.2 Computational Tools for Hard-machining Modelling

6.2.1 Hard-machining Modelling Purposes

Mathematical modelling of hard-machining processes is carried out for two main purposes. On one hand, it is used for obtaining relationships between cutting variables in order to be used in process planning and optimization. These models usually relate cutting parameters (depth of cut, feed, cutting speed, *etc.*) with important process variables, such as cutting temperature, tool life or obtained surface roughness. These relationships are mainly stationary, *i.e.*, they do not explicitly include cutting time.

On the other hand, modelling allows monitoring of the cutting processes, by establishing the relationship between some easy-to-obtain parameters, such as the cutting power or the spindle current, and other relevant variables, like the tool wear. Furthermore, this kind of modelling permits identifying certain values of the measured variables, indicating some important event into the machining process (*e.g.*, cutting tool failure). In both cases, the relationship is transient, that is, it explicitly involves time.

It may be noted that most of the papers published on hard-machining modelling study the turning process [1–7], while only a few deal with other processes like milling [8, 9]. Another important fact is the material studied. The most popular used materials are AISI 52100 steel [1, 3, 5, 7, 10] and AISI D2 steel [2, 6–8, 11, 12], although some other ones have been also reported, for example, AISI 3020 austenitic steel [4], AISI AISI H11 (DIN X38CrMoV5) steel [9] and AISI H13 steel [13, 14].

6.2.2 *Conventional Computational Tools*

Widely used from the very beginning of cutting-process modelling, statistical techniques have proved effective in solving important parts of machining modelling problems, even in hard machining. Several recent works have reported the successful use of regression models for different cutting parameters, mainly surface roughness [5], cutting force [9] and tool life [15]. Some researchers attempt to model more than one variable, such as Davim and Figueira [6], who consider surface roughness, cutting forces and tool flank wear, and Arsecularatne *et al.* [12], who model surface roughness and cutting-force components.

Analysis of variance (ANOVA) has been used for computing the influence of cutting parameters on surface roughness, tool wear and cutting-force components [6, 13]. Furthermore, this technique is widely used in multiple regressions in order to test the validity of the obtained model.

The Taguchi robust method is another reported technique in hard-turning modelling. It has been applied for modelling the effects of cooling on tool wear [16] and to predict tool wear and surface roughness *versus* cutting parameters [17].

Several recent papers [2, 4, 11, 14] compare performance of statistical multiple regressions and artificial neural networks in modelling some variables. They usually claim to have obtained better outcomes by using neural networks than by using conventional statistical tools. However, there is a lack of rigorous techniques for comparing these approaches, therefore, the shortcomings of the statistical approaches are not fully proved, although it is commonly accepted that cutting phenomena in hard turning are still not yet well understood [1].

6.2.3 *Intelligent Techniques*

6.2.3.1 Artificial Neural Networks

Due to the complexity of cutting-process phenomena, there is a heavy nonlinearity in the relationships between the involved variables. For this reason, several researchers have pointed out the shortcomings of the statistical approaches in modelling these relationships [18].

On the contrary, some artificial-intelligence-based tools have proved their ability to match complex nonlinear relationships. The most popular and deeply studied techniques in soft computing are the artificial neural networks. They have been successfully used for modelling different phenomena in hard-machining processes.

Artificial neural networks arose as an attempt to model brain structure and functioning. However, besides any neurological interpretation, they can be considered as a class of general, flexible, nonlinear regression models [19].

The network is composed for several simple units, called neurons, arranged in a certain topology, and connected to each other. Neurons are organized into layers. Depending upon their position, layers are called input layer, hidden layer or output layer. A neural network may contain several hidden layers.

If, in a neural network, neurons are connected only to those in the following layers, it is called a feed-forward network (see Figure 6.1). In this group are included multilayer perceptrons (MLP), radial basis function (RBF) networks and self-organizing maps (SOM).

On the contrary, if recursive or feed-back connections exist between neurons in different layers, the network is called recurrent (see Figure 6.2). Elman and Hopfield networks are typical samples of recurrent topologies.

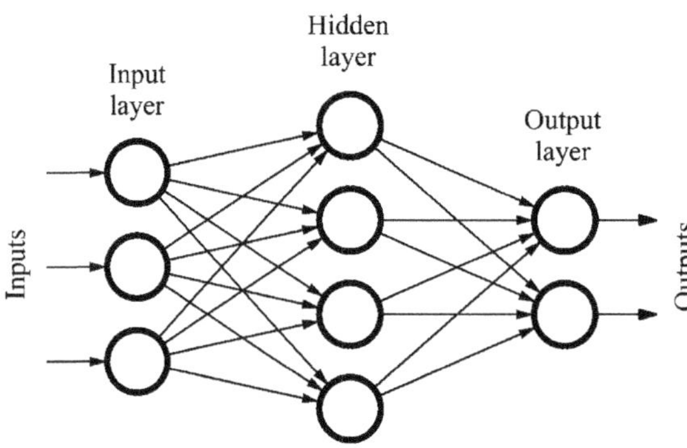

Figure 6.1 Example of a feed-forward neural network

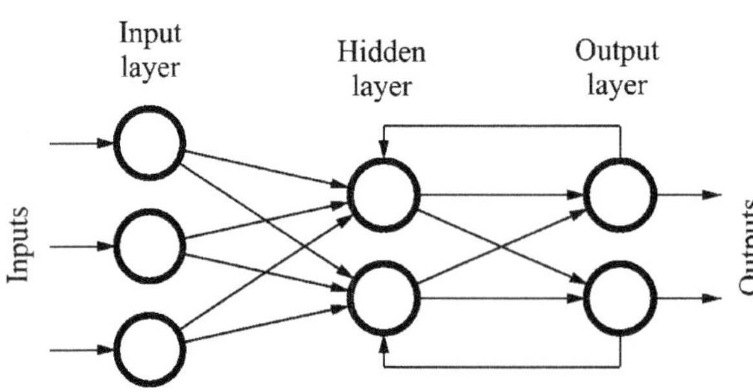

Figure 6.2 Example of a recurrent neural network

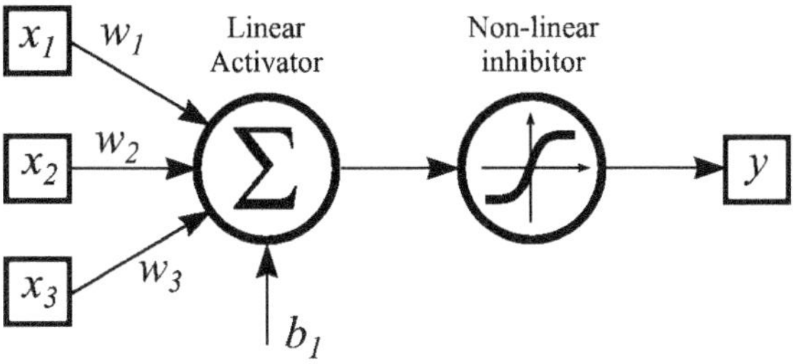

Figure 6.3 Logical scheme of a neuron

A typical neuron consists of a linear activator followed by a nonlinear inhibiting function (see Figure 6.3). The linear activation function yields the sums of weighted inputs plus an independent term so-called bias, *b*.

The nonlinear inhibiting function attempts to arrest the signal level of the sum. Step, sigmoid and hyperbolic tangent functions are the most common functions used as inhibitors (see Figure 6.4). Sometimes, purely linear functions are used for this purpose too, especially in output layers.

The process of adjusting weights and biases, from supplied data, is called training and the used data, training set. The process of training a neural network can be broadly classified into two typical categories:

- Supervised learning: requires using both the input and the target values for each sample in the training set. The most common algorithm in this group is the back-propagation, used in the MLP, but it also includes most of the training methods for recurrent neural networks, time delay neural networks and RBF networks.
- Unsupervised learning: used when the target pattern is not completely known. It includes the methods based on the adaptive resonance theory and SOM.

Back-propagation, which is applied to MLPs, is the most popular and well-studied training algorithm. It is a gradient-descendent method that minimizes the mean-square error of the difference between the network outputs and the targets in the training set.

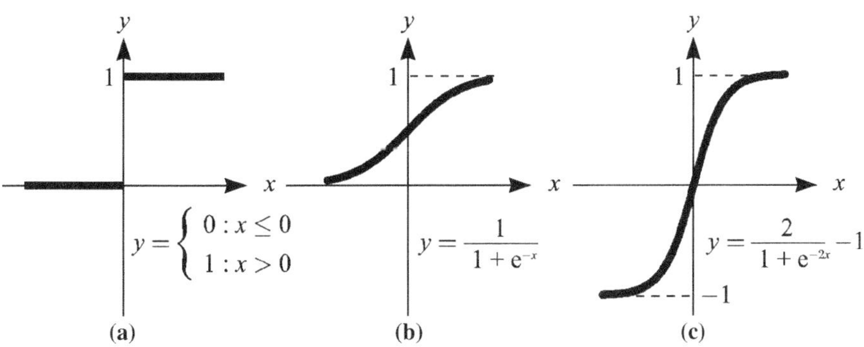

Figure 6.4 Typical inhibiting functions: (a) step, (b) sigmoid, and (c) hyperbolic tangent

Nonlinear function approximation is one of the most important applications of multilayer neural networks. It has been proved that a two-layer neural network can approximate any continuous function, within any arbitrary pre-established error, provided that it has a sufficient number of neurons in the hidden layer. This is the so-called universal approximation property.

In hard machining, artificial neural networks have been widely used, not only for modelling of variables [1, 2, 7, 11], but also for monitoring purposes [20]. A very interesting approach is presented by Umbrello and co-workers [3], who combine neural networks and finite-element methods to predict residual stresses and the optimal cutting conditions during hard turning.

Even for the most widely implemented MLP neural network, there are still no general rules to specify the number of hidden layers, the number of neurons for each layer, and the network connection to achieve an optimized modelling effect. If artificial neural networks are selected as a tool wear modelling approach, such challenges must be carefully addressed [1].

Another drawback is that few papers present the mathematical model of the trained neural network, *i.e.*, the coefficients of weights and biases. This does not allow using the outcomes in other applications.

6.2.3.2 Fuzzy Logic and Neuro-fuzzy Systems

Fuzzy logic, which is based on fuzzy set theory, deals with uncertainty. While binary logic uses only two values for their sets (1 or 0), in fuzzy logic the degree of truth of a statement can range between 0 and 1.

A fuzzy set is a subset of elements, each one having an associated value, from the interval [0, 1] which defines its membership to certain set. These values are also known as degrees of truth, and their distribution is called a membership function.

For example, in Figure 6.5 membership functions for three subsets of cutting force, F_C, are shown. They are called low, moderate and high. Therefore for a force $F_C = 2500$ N, the following statements have the indicated degree of truth:

$$\left.\begin{array}{c} F_C \text{ is "Low"} = 0.20 \\ F_C \text{ is "Moderate"} = 0.80 \\ F_C \text{ is "High"} = 0 \end{array}\right\} \tag{6.1}$$

A general fuzzy inference system consists of three parts. A crisp input is firstly fuzzified by expressing the input variables in the form of fuzzy membership values based on various membership functions. Then, a fuzzy rule base processes it to obtain a fuzzy output. Finally, this fuzzy output is defuzzified to give a crisp outcome.

Because of the complementary nature of fuzzy logic and neural networks, these two techniques can be integrated in a number of ways to overcome the drawbacks of each. Such connectionist architectures are commonly known as neuro-fuzzy

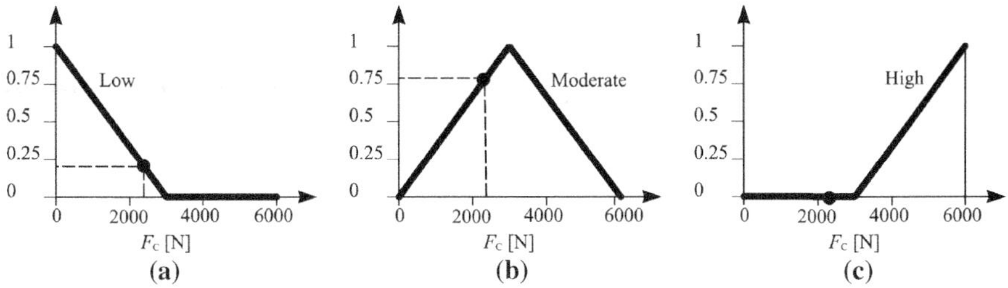

Figure 6.5 Samples of membership functions: (a) low, (b) moderate, and (c) high

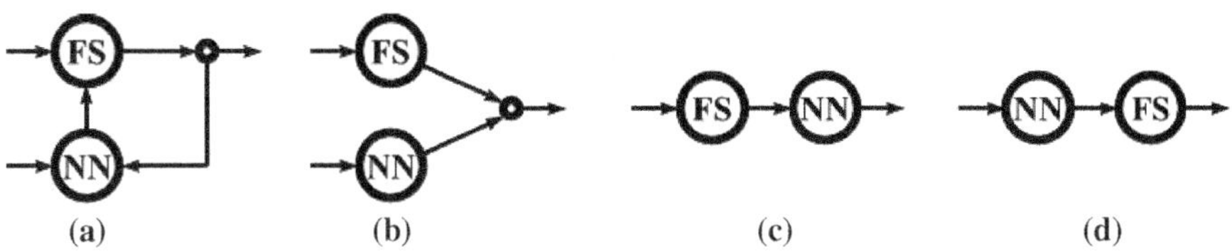

Figure 6.6 Combinations of neural networks and fuzzy logic: (a) NN optimizes the parameters of FS, (b) output of FS is corrected by the output of NN, and (c), (d) cascade assemblies

hybrid systems [21]. While neural networks support the system to have the capability of training from empirical data, fuzzy logic provides reasoning for the training process and generates a rule bank for control or classification purposes.

Although there are many possible combinations of the two systems, there are four basic combinations that have been successfully applied (see Figure 6.6). The first combination (Figure 6.6 (a)) uses the neural network (NN) to optimize the parameters of the fuzzy system (FS) by minimizing the gap between the output of the fuzzy system and the given target. In the second one (Figure 6.6 (b)) the output of a fuzzy system is corrected by the output of a neural network to increase the precision of the final system output. The third and fourth combinations (Figure 6.6 (c) and (d)) are cascade assemblies of a neural network and a fuzzy system.

The set of published works on neuro-fuzzy system applications to hard turning include the paper of Horng and Chiang [22] who use fuzzy logic for modelling tool wear and surface roughness in turning Hadfield steel. In another approach Huang and Chen implement a fuzzy-nets-based in-process surface roughness prediction system for turning operations [23].

In spite of some successful applications, neuro-fuzzy systems are not very simple to implement, and their models are difficult to used with other systems.

6.2.3.3 Other Intelligent Tools

Support vector machines can take advantage of prior knowledge and construct a hyperplane as the decision surface so that the margin of the separation between different classes is maximized. The support vector machine was initially developed for the classification problem with separable data, and later it was improved to handle non-separable data.

Support vector machines have been successfully applied [24] for multiclassification of tool wear in a turning process.

6.3 Optimization of Hard Machining

6.3.1 Importance of Hard-machining Optimization

Optimization is an important task in machining processes, allowing selection of the most convenient cutting conditions in order to obtain desired values in some variable, which usually has a direct economical impact, such as machining time or total operation cost.

Optimization of machining processes is usually difficult, where the following aspects are required:

- knowledge of cutting process;
- empirical equations relating the tool life, forces, power, *etc.*, for developing realistic constraints;
- specification of machining capabilities;
- development of an effective optimization criterion; and
- knowledge of mathematical and numerical optimization techniques.

In hard machining, optimization tasks are critical, because it involves many complex processes. Usually, small variation in one parameter causes notable changes in other one. Moreover, some variables, such as cutting forces or tool wear, heavily depend upon the cutting conditions.

Therefore the optimization of hard-machining processes is not fully solved yet. In the following sections, the main points on this topic are reviewed and explained, taking into account the most recent publications in this field.

6.3.2 Problem Definition

6.3.2.1 Single-objective Optimization

Single-objective optimization can be viewed as the problem of finding a vector of decision variables, **x**, which satisfies constraints and optimizes a scalar objective function, y. Hence, the term "optimize" means finding a minimum or maximum value; however, it is possible to deal only with minimization problems because any maximization problem can be turned into a minimization one by conveniently transforming the objective function.

In a more formal way, a single-objective optimization problem can be defined as follows:

Definition 6.1 (*single-objective optimization problem*). Given a scalar function $y : \Omega \subset \mathbf{R}^n \to \mathbf{R}, \Omega \neq \varnothing$, find the value $\mathbf{x}^* \in \Omega$ (called the global minimum solution), which minimizes (or maximizes) the value of y, *i.e.*, $\neg\exists \mathbf{x}' \in \Omega : y(\mathbf{x}') < y(\mathbf{x})$.

The set Ω is the feasible region, which is usually defined as:

$$\Omega = \{\mathbf{x} \in \mathbf{R}^n \mid (g_i(\mathbf{x}) \geq 0, i = 1, \ldots, m) \wedge (h_i(\mathbf{x}) = 0, i = 1 \ldots, p)\}$$

where $g_i(\mathbf{x})$ are the m inequality constraints and $h_i(\mathbf{x})$ are the p equality constraints.

6.3.2.2 Multi-objective Optimization

Roughly speaking, multi-objective optimization can be considered as the problem of simultaneously minimizing (or maximizing) two or more target functions. In a more formal way:

Definition 6.2 (*multi-objective optimization problem*). Given the vector function $\mathbf{y} : \Omega \subset \mathbf{R}^n \to \mathbf{R}^k, \Omega \neq \varnothing$, find the value of $\mathbf{x} \in \Omega$ that minimizes (or maximizes) the components of the vector $\mathbf{y}$.

As in a single-objective optimization problem, Ω is restricted by inequality and equality constraints.

However, there is not a formal criterion for comparing two vectors, so a global minimum may not exist. In this sense, two main approaches can be used. The first one is the *a priori* technique, where the decision maker combines the different objectives into a scalar cost function. This actually turns the multi-objective problem into a single-objective one, before the optimization process is carried out. Into this approach are included the linear and nonlinear combination and the aggregation by ordering.

The second approach is called *a posteriori*. In this technique, the decision maker is presented with a set of optimal candidate solutions and chooses from that set. These solutions are optimal in the wide sense that no other solution in the search space is superior to them when all the optimization objectives are considered.

Therefore, in order to formalize the *a posteriori* approach for the multi-objective optimization problem, some preliminary definitions must be made:

Definition 6.3 (*Pareto dominance*). A vector, $\mathbf{u} = (u_1, \ldots, u_k)$, is said to dominate another vector, $\mathbf{v} = (v_1, \ldots, v_k)$ (denoted by $\mathbf{u} \prec \mathbf{v}$), if and only if no component of $\mathbf{u}$ is greater than the corresponding component of $\mathbf{v}$, and at least one component of $\mathbf{u}$ is smaller; *i.e.*:

$$\mathbf{u} \prec \mathbf{v} : (\mathbf{u}, \mathbf{v}) \in \mathbf{R}^k \Leftrightarrow (\forall i \in \{1, \ldots, k\}, u_i \leq v_i) \wedge (\exists i \in \{1, \ldots, k\} : u_i < v_i).$$

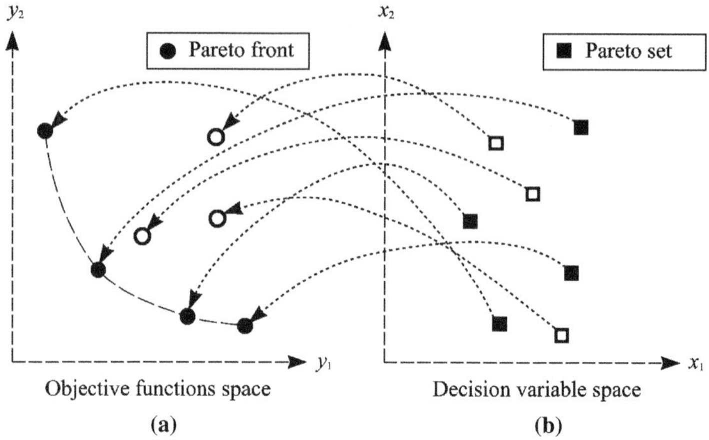

Figure 6.7 Graphical representation of the Pareto front (a) and the Pareto set (b)

Definition 6.4 (*Pareto optimality*). A solution, $\mathbf{x} \in \Omega$, is said to be Pareto optimal with respect to Ω, (denoted as $\mathbf{x} = \text{par}(\Omega)$) if and only if there is not $\mathbf{x}' \in \Omega$ for which $\mathbf{y}(\mathbf{x}')$ dominates $\mathbf{y}(\mathbf{x})$.

Pareto-optimal solutions are also termed non-inferior, admissible, or efficient solutions.

Definition 6.5 (*Pareto-optimal set*). The Pareto-optimal set (denoted by P) is defined as set of all the Pareto optimal solutions, for a given multi-objective optimization problem, *i.e.*:

$$P \equiv \{\mathbf{x} \in \Omega \mid \neg\exists \mathbf{x}' \in \Omega : \mathbf{y}(\mathbf{x}') \prec \mathbf{y}(\mathbf{x})\}.$$

Definition 6.6 (*Pareto front*): The Pareto front (denoted by F) is defined as the set of the images $\mathbf{y}(\mathbf{x})$ for all the values of the Pareto-optimal set, *i.e.*:

$$F \equiv \{\mathbf{y} = \mathbf{y}(\mathbf{x}) \mid \mathbf{x} \in P\}.$$

In Figure 6.7 is shown a graphical representation of the Pareto set and the Pareto front for a two-dimensional function of a two-dimensional argument.

Finally, a multi-optimization problem, in the *a posteriori* approach, can be defined as the problem of finding the Pareto set.

6.3.3 Objective Function

Selection of a proper objective function is very important in setting up the optimization problem. The selected criterion must reflect the most relevant target, taking into account the characteristics of the considered process.

Like in other machining processes, in hard machining, the most common objective function is the cost, because it has a clear direct economical meaning. Several papers report the use of the cost as an optimization criterion; some of them consider only the labour cost, Z_L, which is a function of the machining time, τ. On the contrary, some others prefer to consider a combined cost, Z, that includes not only labour cost but also overhead, Z_O, and tool costs, Z_T [25].

Another very popular optimization objective is the machining time, τ. It has a heavy influence on the economy of the process, especially in these cases where the tool cost can be neglected when comparing with labour and overhead cost. On the other hand, material removal rate, as the inverse magnitude of the machining time, is also used as an optimization target [26].

Xueping and co-workers [27] have reported the optimization of residual stress in hardened bearing steel.

Nevertheless, these single-objective approaches have a limited value in fixing the optimal cutting conditions, due to the complex nature of the hard-machining processes, where several different and contradictory targets must be simultaneously considered.

Currently multi-objective methods are the most popular approaches in hard-machining optimization and they have been widely reported in the specialized literature. Combinations of time and cost [7], tool wear and surface roughness [17, 28] and time and roughness [29] have been carried out.

Bouacha *et al.* [30] present a combination of six objective functions: three measures for the surface roughness (R_a, R_s and R_z) and the three components of the cutting force, F_C, F_F and F_R. Another interesting approach is given by Paiva and co-workers [31], who optimize simultaneously tool life, T, processing cost per piece, C_p, cutting time, τ, the total turning cycle time, τ_T, surface roughness, R_a, and the material removing rate.

It must de noted that most of the works mentioned use *a priori* approaches. Only Özel and Karpat [29] obtain the Pareto front for their combination of objective function.

6.3.4 Decision Variables

Usually, the decision variables in hard-machining optimization problems include the cutting parameters. In hard turning sometimes only the feed rate, f, and the cutting speed, v, are considered [29] but in other cases, the depth of cut, a_P, is also included [17, 27, 30, 31]. Basak *et al.* [7] consider the cutting time as another decision variable.

Occasionally, other parameters are considered as decision variables, reflecting some important aspects of the problem. In this group are included the tool geometry, reflected by the nose radius, r_E [26, 28] or tool diameter [25, 32].

6.3.5 Constraints

Although they are not formally constraints, the valid limits for the decision variables are considered in the delimitation of the feasible region. For example, for the cutting parameters, they take the form:

$$a_P^{min} \leq a_P \leq a_P^{max}, \tag{6.2a}$$

$$f^{min} \leq f \leq f^{max}, \tag{6.2b}$$

$$v^{min} \leq v \leq v^{max}, \tag{6.2c}$$

where $[a_P^{min}, a_P^{max}]$, $[f^{min}, f^{max}]$ and $[v^{min}, v^{max}]$ are the valid ranges for the respective variables.

Other constraints usually taken into account are the cutting and feed forces, F_C and F_F, which must be less than the allowed values for the machine tool, F_C^{max} and F_F^{max}:

$$F_C \leq F_C^{max}, \tag{6.3a}$$

$$F_F \leq F_F^{max}. \tag{6.3b}$$

In finishing passes, the surface roughness, R, is usually included as a constraint, keeping it below the pre-established value R^{max} [7, 26]:

$$R \leq R^{max}. \tag{6.4}$$

Paiva and co-workers [31] propose a combination of the objective functions as a constraint, in order to give another set of priorities for these objectives.

6.3.6 Optimization Techniques

6.3.6.1 General Considerations

Optimization techniques can be grouped into two broad categories: numeric and stochastic approaches. The first group comprises exact algorithmic methods, with a solid mathematical foundation, for obtaining the global optimum. Numeric approaches commonly use iterative algorithms. They include the gradient-based approaches, the descendent method and the simplex algorithm.

On the contrary, stochastic optimization tries to imitate some natural processes, which, although they do not guarantee the consecution of the global optimum, they allow good enough solutions to be obtained. These heuristics have a strong random component. Techniques such as evolutionary algorithms, simulated annealing and particle swarm optimization are included in this group.

6.3.6.2 Response Surface Methodology

Response surface methodology is a general approach for obtaining the maximum value of a dependent (response) variable which depends upon several independent (explanatory) variables. This technique combines design of experiments (DoE) and multiple regression.

DoE is a general approach for designing any information-gathering exercises where variation is present. In machining-process modelling, DoE deals mainly with controlled experiments, where variations in the independent variables are under the control of the researcher.

Current DoE techniques are based in the following principles:

- Comparison: helps detecting undesirable variation in the measured results.
- Randomization: decreases the influence of uncontrolled factors on the measured results.
- Replication: allows estimating the variation in the measured results.
- Blocking: reduces known but irrelevant sources of variation between units and, therefore, increases precision in the estimation of the source of variation.
- Orthogonality: assures that contrast will be carried out in such a way that all the information can be captured.

The most common approach in DoE are factorial experiments, which provide a suitable distribution of experimental points, according to the mentioned principles. In Figure 6.8 some samples of DoE are shown.

Several papers have reported using response surface methodology (either the complete methodology or some of its components) for optimizing hard-machining processes [25, 31, 32].

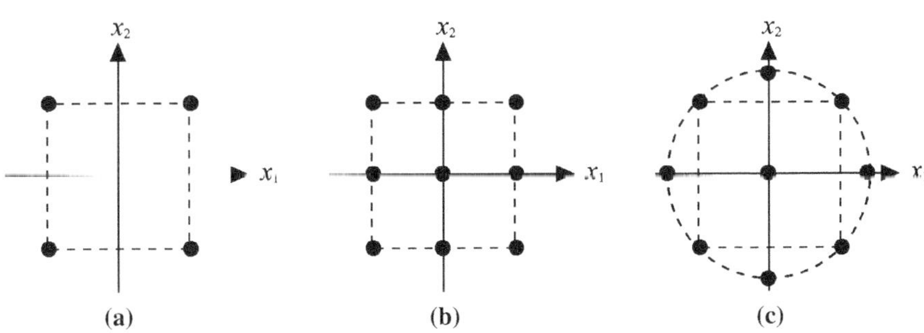

Figure 6.8 Examples of factorial designs: (a) full factorial 2^2, (b) full factorial 3^2, and (c) cube-star

6.3.6.3 Taguchi Method

The Taguchi method, also known as Taguchi's robust design, is a methodology for designing products that have minimum sensitivity to variation in uncontrollable factors. This method is based on conventional techniques, with some guidelines for laying out design experiments and analysing the outcomes of these experiments.

A key role in the Taguchi method is played by the so-called quality loss function, which is a continuous function that is defined in terms of the deviation of a design parameter from an ideal or target value.

The loss function is assumed to be quadratic (see Figure 6.9), so it can be expressed by the equation:

$$L(y) = k(y - m)^2, \tag{6.5}$$

where m is the actual value for the design parameter, y is its actual value, and k is a constant that depends on the cost at the specification limits. For a group of experimental data, with an average value of μ, and a variance of σ^2, the average quality loss function can be estimated as:

$$L(y) = k[S^2 + (\mu - m)^2]. \tag{6.6}$$

In parameter design, there are two types of factors that affect a product's functional characteristic: control factors and noise factors. Control factors are those factors which can easily be controlled; on the contrary, noise factors are those that are difficult or impossible or too expensive to control.

In order to vary and test the different levels of the control factor, a special DoE technique, called orthogonal arrays, is used. Actually, two arrays, containing the control factors and noise factors, are combined. They are named inner array and outer array, respectively. Together they are known as a product array or complete parameter design layout.

The most convenient parameter settings are computed from the experimental data, by using the signal-to-noise ratio, SN. Levels that maximize the desired sig-

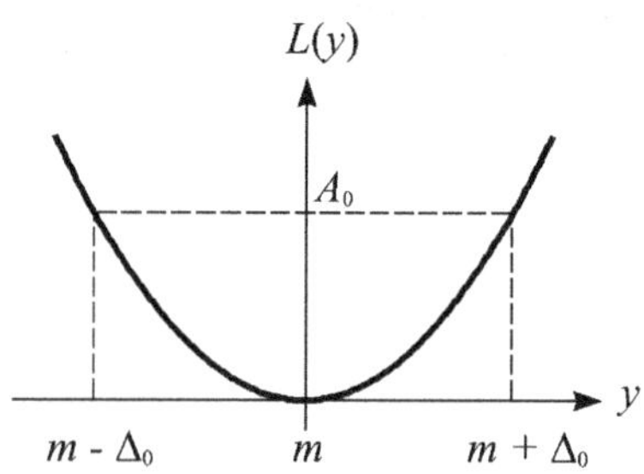

Figure 6.9 Samples of quality loss functions

nal-to-noise ratio are considered as optimal. There are three standard signal-to-noise ratios, depending on the desired performance response:

- smaller the better (for making the system response as small as possible):

$$SN = -10\log\left(\frac{1}{n}\sum_i y_i^2\right); \tag{6.7a}$$

- nominal the best (for reducing variability around a target):

$$SN = 10\log\left(\frac{\bar{y}^2}{\sigma^2}\right); \tag{6.7b}$$

- larger the better (for making the system response as large as possible):

$$SN = -10\log\left(\frac{1}{n}\sum_i \frac{1}{y_i^2}\right). \tag{6.7c}$$

Once all *SN* have been computed for each run of an experiment, a graphical approach is used to analyze the data. In this graphical approach, *SN* and average responses are plotted for each factor against each of its levels. The graphs are then examined to select the factor level which best maximizes *SN* and brings the mean on target (or maximizes or minimizes the mean, as the case may be).

Several applications of Taguchi method have been reported for optimizing hard machining, not only for single objective applications [27] but also for multi-objective ones [17, 28, 30].

6.3.6.4 Evolutionary Algorithms

Evolutionary algorithms are a set of heuristics simulating the process of natural evolution (Figure 6.10). Although the underlying mechanisms are simple, these algorithms have proven them as a general, robust and powerful search tool. In particular, they are especially convenient for problems involving multiple conflicting objectives and large and complex search spaces.

In spite of the wide diversity in the proposed approaches, an evolutionary algorithm can be characterized by three features:

- A set of candidate solutions is maintained.
- A competitive selection process is performed on this set.
- Several solutions may be combined in terms of recombination to generate new solutions.

There are two main evolutionary heuristics: the German school of evolution strategies (ES), and the American school of GAs. The main differences between these two approaches are summarized in Table 6.1.

Some papers have shown the application of evolutionary techniques in hard-machining optimization [26, 33].

Figure 6.10 Block diagram of a typical evolutionary algorithm

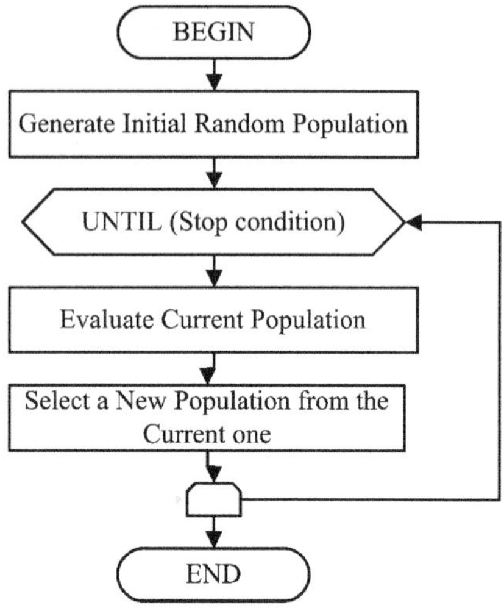

Table 6.1 Differences between ES and GA

Characteristic	ES	GA
Individual codification	As real number	As bitstring
Mutation	By adding a random number	By changing a bit
Crossover	By combining real numbers	By breaking and recombining bitstring

6.4 Case Study

6.4.1 Case Description

In this case study, modelling and optimization of a hard-turning process is presented. Experimental data was taken from a paper by Davim and Figueira [34], on a turning process of a high-chromium cold-work tool steel AISI D2 steel (1.55 % C, 0.30 % Si, 0.40 % Mn, 11.80 % Cr, 0.80 % Mo and 0.80 % V). Workpieces were hardened by quenching (after vacuum treatment) between 1000 and 1040 °C, achieving a hardness of ~59 HRC.

Ceramic tool inserts of type CNMA 120408 T01020 CC650 were used to machine the tool steel with a geometry as follows: rake angle –6° (negative), clearance angle 5°, edge major tool cutting inclination angle 80° and cutting-edge inclination angle 0°. A tool holder of type PCLNL2020K12 (ISO) was used. The depth of cut, a_P, was 0.2 mm.

Three independent variables were considered in the experimental design: cutting speed, v, feed, f, and cutting time, τ. In Table 6.2 the corresponding experimental levels for each variable are shown.

A full factorial experimental design 3^3 was carried out. Values of the specific cutting pressure, K_S, the tool wear, V_C, and the arithmetic mean of the surface roughness, R_a, were measured for each point. Experimental results are shown in Table 6.3.

Table 6.2 Experimental levels for the independent variables [34]

Variable	Low level	Middle level	High level
Cutting speed, v (m/min)	80	150	220
Feed, f (mm/rev)	0.05	0.10	0.15
Time, τ (min)	5	10	15

Table 6.3 Experimental data [34]

No.	Cutting speed, v (m/min)	Feed, f (mm/rev)	Time, t (min)	Specific pressure, K_S (N/mm^2)	Tool wear, V_C (mm)	Roughness, R_a (μm)
1	80	0.05	5	4264.6	0.058	0.43
2	80	0.05	10	4127.1	0.081	0.50
3	80	0.05	15	3775.9	0.104	0.56
4	80	0.10	5	2872.2	0.048	0.72
5	80	0.10	10	3137.8	0.080	0.97
6	80	0.10	15	3218.8	0.088	1.05
7	80	0.15	5	2320.8	0.033	0.55
8	80	0.15	10	2537.0	0.081	0.70
9	80	0.15	15	2671.4	0.960	0.82
10	150	0.05	5	3808.0	0.101	0.39
11	150	0.05	10	4172.4	0.140	0.50
12	150	0.05	15	4290.2	0.540	0.60
13	150	0.10	5	2843.9	0.082	0.79
14	150	0.10	10	3085.5	0.141	0.89
15	150	0.10	15	3654.7	0.250	1.05
16	150	0.15	5	2497.5	0.099	0.77
17	150	0.15	10	2661.4	0.176	1.07
18	150	0.15	15	3131.0	0.243	1.32
19	220	0.05	5	4783.7	0.339	0.26
20	220	0.05	10	5037.1	0.597	0.46
21	220	0.05	15	5590.4	0.642	1.40
22	220	0.10	5	3126.6	0.200	0.59
23	220	0.10	10	3756.3	0.300	0.98
24	220	0.10	15	4386.9	0.320	1.38
25	220	0.15	5	2655.8	0.188	0.88
26	220	0.15	10	3134.2	0.225	1.31
27	220	0.15	15	3109.1	0.775	1.48

6.4.2 Statistical Modelling

To establish a useful relationship between independent variables (tool wear, surface roughness and specific cutting force) and dependent variables (cutting speed, feed and machining time), multiple regression models were adjusted. Three types of equations were tried for each model: linear, quadratic and potential.

For specific pressure, K_S, the following models were obtained:

$$K_S = 3877 + 5.281v - 16.81 \cdot 10^3 f + 51.73t, \quad R^2 = 0.88, \quad p(F) = 0.0000; \tag{6.8a}$$

$$\begin{aligned} K_S = {} & 7204 - 18.62v - 44.59 \cdot 10^3 f - 166.7t - 8.519vft + 40.58vf \ldots \\ & \ldots + 1.402vt + 1490ft + 47.81 \cdot 10^{-3} v^2 + 97.83 \cdot 10^3 f^2 - 0.658t^2, \\ & R^2 = 0.96, \quad p(F) = 0.0000; \end{aligned} \tag{6.8b}$$

$$K_S = 363.7 \frac{v^{0.1832} t^{0.1382}}{f^{0.4299}}, \quad R^2 = 0.91, \quad p(F) = 0.0000. \tag{6.8c}$$

As can be seen, all the models as fitted explain more than 85 % of the variability in K_S. However, the quadratic model is the best one. All of them show a statistically significant relationship between the independent and dependent variables, with more than 99 % of confidence level.

For tool wear, V_C, the following models were adjusted:

$$V_C = -0.317 + 1.63 \cdot 10^{-3} v + 0.198f + 0.0308t, \quad R^2 = 0.41, \quad p(F) = 0.0038; \tag{6.9a}$$

$$\begin{aligned} V_C = {} & 1.22 - 4.26 \cdot 10^{-3} fvt - 5.08 \cdot 10^{-3} v - 15.2f - 0.126t + \ldots \\ & \ldots + 13.6 \cdot 10^{-3} fv + 0.929ft + 0.426 \cdot 10^{-3} vt + 52.5f^2 - \ldots \\ & \ldots - 17.9 \cdot 10^{-6} v^2 + 3.17 \cdot 10^{-3} t^2, \\ & R^2 = 0.71, \quad p(F) = 0.0073; \end{aligned} \tag{6.9b}$$

$$V_C = 23.4 \cdot 10^{-6} \frac{v^{1.30} t^{1.05}}{f^{0.0839}}, \quad R^2 = 0.69, \quad p(F) = 0.0000. \tag{6.9c}$$

Evidently, the linear model has a poor performance. Although the other ones are slightly better, their correlation is actually insufficient. The three models show a probability value, associated to the Fisher statistic, lower than 0.01, so there is a statistically significant relationship, at the 99 % confidence level.

Finally, for the surface roughness, R_a, the obtained models were:

$$R_a = -0.357 + 1.94 \cdot 10^{-3} v + 4.22f + 0.0476t, \quad R^2 = 0.70, \quad p(F) = 0.0000; \tag{6.10a}$$

$$R_a = 0.927 - 10.5 \cdot 10^{-3} v - 6.29 f - 0.100 t + 4.86 \cdot 10^{-3} vft + \ldots$$
$$\ldots + 71.7 \cdot 10^{-3} vf + 0.914 \cdot 10^{-3} vt + 0.709 ft + 3.17 \cdot 10^{-6} v^2 - \ldots$$
$$\ldots - 63.1 f^2 + 0.622 \cdot 10^{-3} t^2, \tag{6.10b}$$
$$R^2 = 0.89, \quad p(F) = 0.0000;$$

$$R_a = 0.310 v^{0.229} t^{0.532} f^{0.588}, \qquad R^2 = 0.73, \quad p(F) = 0.0000. \tag{6.10c}$$

The quadratic model shows the best performance. Nevertheless, none of them explains more than 90 % of the variability in the experimental data. The three models exhibit a statistically significant relationship, at the 99 % confidence level.

6.4.3 *Neural-network-based Modelling*

6.4.3.1 General Aspects

To establish a useful relationship between independent variables (tool wear, surface roughness and specific cutting force) and dependent variables (cutting speed, feed and machining time), MLP-type neural networks were selected. The neural networks have two layers: one hidden layer and one output layer. The hidden layer uses a sigmoid-type transference function:

$$f(\mathbf{x}) = \frac{1}{1 + \exp(-b - \sum w_i x_i)}; \tag{6.11a}$$

while the output layer uses a linear function:

$$f(\mathbf{x}) = b + \sum w_i x_i \,. \tag{6.11b}$$

To carry out the training process, not only input variables but also output ones were normalized in the range [0, 1] in order to facilitate the neural-network training process.

The networks were trained by using the gradient descendent with adaptive velocity and momentum back-propagation algorithm. The learning rate was established as 0.01; the learning rate increase, 1.05; learning rate decrease, 0.9; momentum constant, 0.7; maximum error ratio, 1.04; and maximum number of epochs to train, 2000. When created, all the weights and biases were randomly initialized in the range –0.1 ... 0.1.

6.4.3.2 Specific Cutting-force Model

For the specific cutting force, K_S, three neurons were established in the hidden layer. This guarantees the existence of enough degrees of freedom (as can be noted in Table 6.4), for making the training process mathematically determined.

The pre-established training error was achieved at epoch number 4915. In Figure 6.11 is shown the sum of squared errors (SSE), as the training process is carried out.

The *R*-squared statistic for this model was 0.96, which indicates that the model as fitted explains 96 % of the variability in K_S. Since the *P*-value in the ANOVA table (Table 6.4) is less than 0.01, there is a statistically significant relationship between the variables at the 99 % confidence level.

Table 6.4 ANOVA for the K_S model

Source	Sum of squares	D.F.	Mean squares	*F*-ratio	*P*-value
Model	17.988×10^6	16	1.124×10^6	21.04	0.0000
Residual	0.053×10^6	10	0.053×10^6	–	–
Total (corr.)	18.675×10^6	26	–	–	–

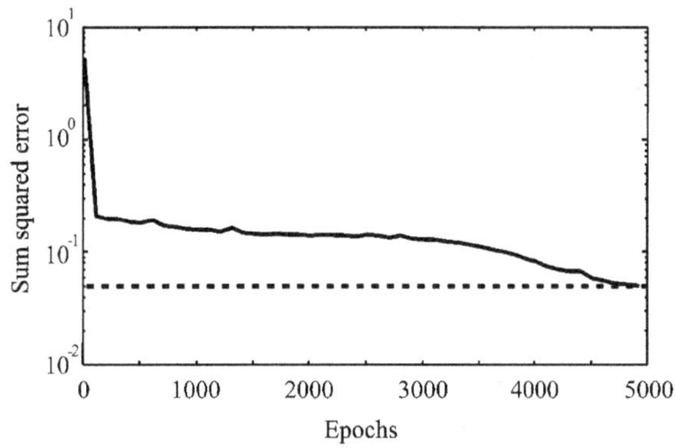

Figure 6.11 Training process for the K_S network

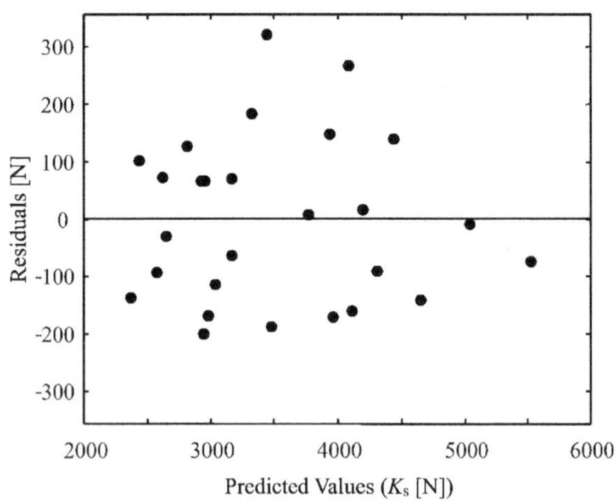

Figure 6.12 Residual *vs.* predicted values for the K_S model

In Figure 6.12 are shown the residual plotted *versus* the predicted values by the model. It can be noted that residuals are homogeneously distributed. In Code 6.1 is listed the function for the neural network based model of K_S.

In Figure 6.13 the graphical representation of the adjusted model is shown. It must be remarked that the relationship between variables is complex, which makes the application of neural networks very convenient.

Code 6.1 MATLAB function of the neural network model for K_S

```
function Ks = nn_ks(v, f, t)
 v = (v - 80)./(220 - 80);
 f = (f - 0.05)./(0.15 - 0.05);
 t = (t - 5)./(15 - 5);
 X = [v; f; t];
 W1 = [-0.738005, -0.366224,  0.916024;
  -0.736798, -2.463920, -0.927616;
   2.039850, -1.271836,  0.624422];
 B1 = [0.638930; -1.197159; -2.863192];
 W2 = [0.597236,  1.888979,  1.877621];
 B2 = [-0.384430];
 Y1 = logsig(W1*X + B1*ones(1,size(X,2)));
 Y2 = purelin(W2*Y1 + B2*ones(1,size(Y1,2)));
 Ks = 2320.8 + (Y2.').*(5590.4 - 2320.8);
```

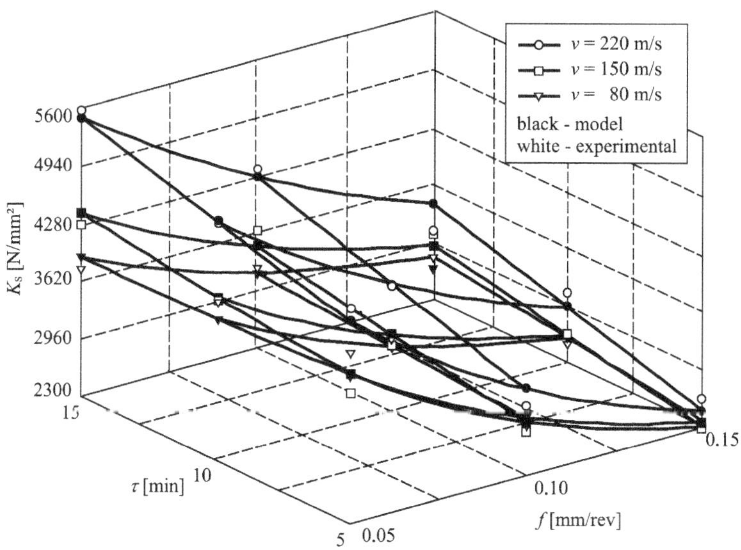

Figure 6.13 Graphical representation of the K_S model

6.4.3.3 Tool Wear Model

For tool wear model, V_C, a network with four neurons in the hidden layer was selected. The SSE was pre-established as 0.10, and was reached at epoch 23,097. In Figure 6.14 the SSE along the training process, is plotted.

R-squared for this model is 0.94, which means that it explains 94 % of the variability of V_C. The ANOVA (see Table 6.5), shows a P-value less than 0.1, so it is possible to say that, at a 90 % confidence level, there is a statistically significant relationship between the studied variables.

Residuals *versus* predicted values are shown in Figure 6.15, showing that residuals are homogeneously distributed. In Code 6.2 it is listed the function for the neural-network-based model of V_C.

The graphical representation of the adjusted model is given in Figure 6.16. The obtained model does not fit completely the experimental data; however, it is better than the correspondent statistical models.

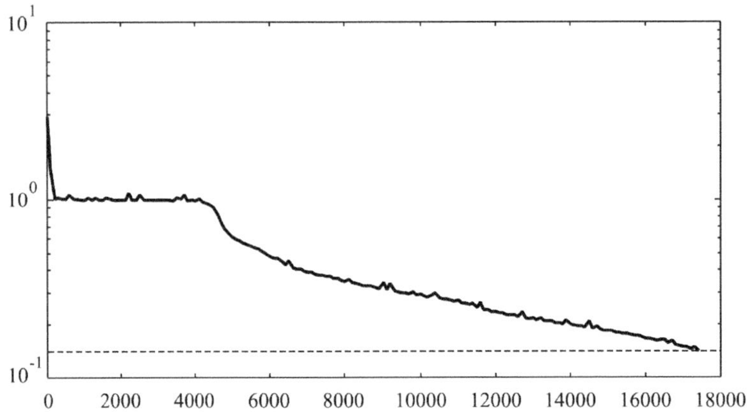

Figure 6.14 Training process for the V_C network

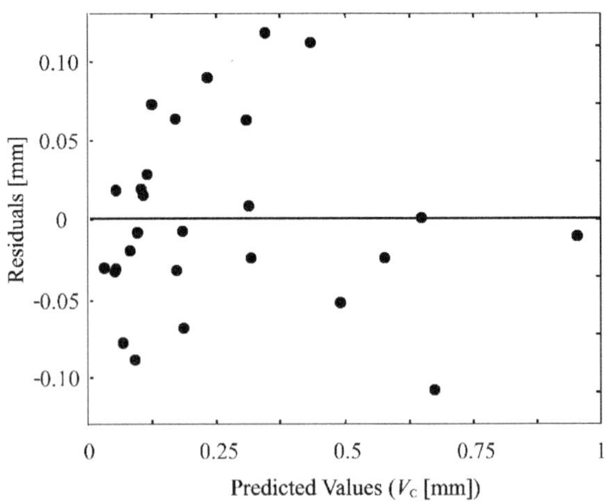

Figure 6.15 Residual *vs.* predicted values for the V_C model

Table 6.5 ANOVA for the V_C model

Source	Sum of squares	D.F.	Mean squares	F-ratio	P-value
Model	1.4198	21	0.0676	3.9341	0.0671
Residual	0.0859	5	0.0172	–	–
Total (corr.)	1.5205	26	–	–	–

Code 6.2 MATLAB function of the neural network model for V_C

```
function Vc = nn_vc(v, f, t)
 v = (v - 80)./(220 - 80);
 f = (f - 0.05)./(0.15 - 0.05);
 t = (t - 5)./(15 - 5);
 X = [v; f; t];
 W1 = [0.895355, 4.197258, 0.926702;
  0.444275, 2.139901, 0.225616;
   -5.100363, 4.328508, 4.664264;
  4.330807, -2.101679, 3.694703];
 B1 = [-3.321501; -2.334190; -10.343271; ...
   -4.972819];
 W2 = [-1.857116, 3.478536, 5.082896, 0.729469];
 B2 = [-0.253010];
 Y1 = logsig(W1*X + B1*ones(1,size(X,2)));
 Y2 = purelin(W2*Y1 + B2*ones(1,size(Y1,2)));
 Y = Y2;
 Vc = 0.033 + (Y2.').*(0.96 - 0.033);
```

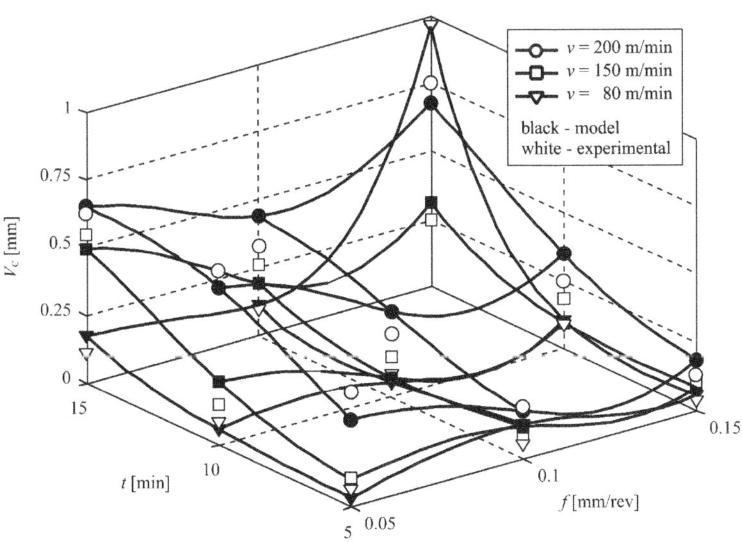

Figure 6.16 Graphical representation of the V_C model

6.4.3.4 Surface Roughness Model

For surface roughness, R_a, three hidden neurons were included in the correspondent network. The expected SSE was established in 0.05, and it was achieved at epoch 20,000. In Figure 6.17, the training process, with corresponding SSEs, is represented.

The adjusted model has a R^2 statistic of 0.97, explaining 97 % of the variability of the model. The ANOVA is shown in Table 6.6. The probability associated with the F statistic is near zero, therefore, with a confidence level of 99 %, there is a statistically significant relationship between the analyzed variables.

In Figure 6.18, residuals are plotted *versus* predicted values, for the R_a model. These residuals are normally distributed and no tendency can be identified from them. The obtained function is listed in Code 6.3. In Figure 6.19, the adjusted model for R_a is graphically shown.

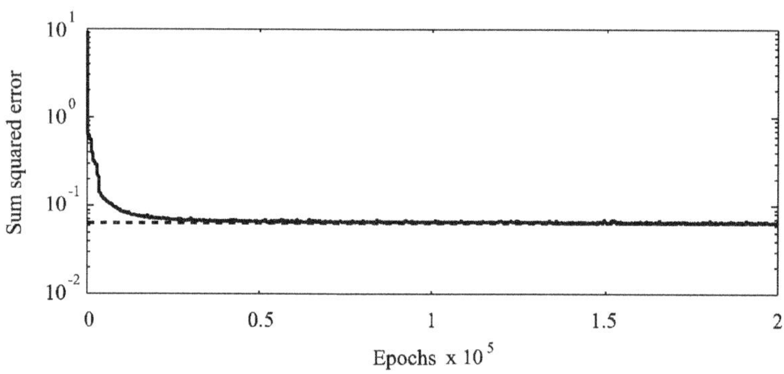

Figure 6.17 Training process for the R_a network

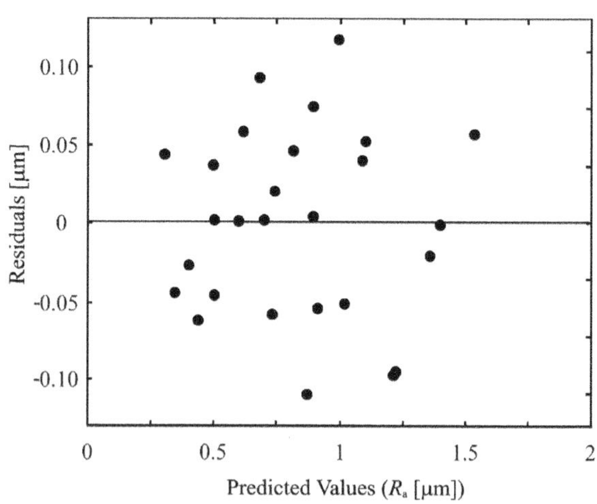

Figure 6.18 Residual *vs.* predicted values for the R_a model

Table 6.6 ANOVA for the R_a model

Source	Sum of squares	D.F.	Mean squares	F-ratio	P-value
Model	2.9548	16	0.1847	19.73	0.0000
Residual	0.0936	10	0.0093	–	–
Total (corr.)	3.0523	26	–	–	–

Code 6.3 MATLAB function of the neural network model for R_a

```
function Ra = nn_ra(v, f, t)
 v = (v - 80)./(220 - 80);
 f = (f - 0.05)./(0.15 - 0.05);
 t = (t - 5)./(15 - 5);
 X = [v; f; t];
 W1 = [-0.442358,  2.072393,  0.385718;
  -1.045139,  3.257369, -0.015000;
   5.368294, -2.052004,  4.133462];
 B1 = [-1.853852; -3.998309; -11.184745];
 W2 = [3.314839, -4.285657,  5.030268];
 B2 = [-0.254177];
 Y1 = logsig(W1*X + B1*ones(1,size(X,2)));
 Y2 = purelin(W2*Y1 + B2*ones(1,size(Y1,2)));
 Y = Y2;
 Ra = 0.26 + (Y2.').*(1.48 - 0.26);
```

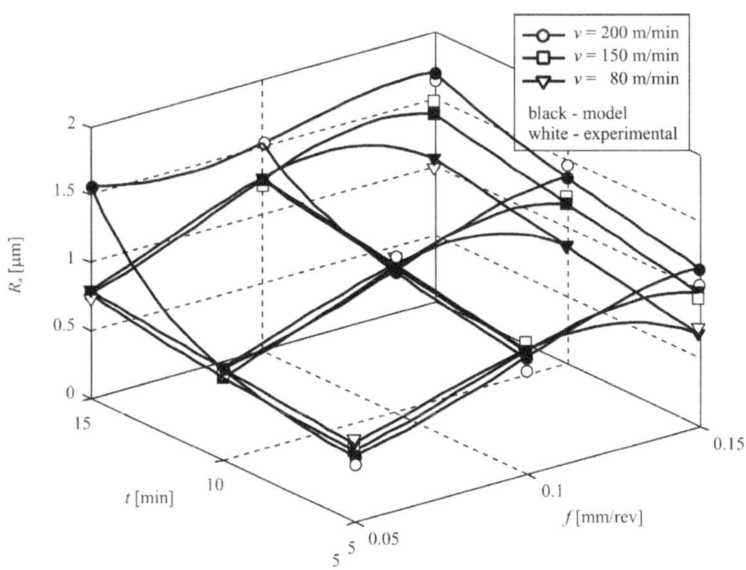

Figure 6.19 Graphical representation of the R_a model

6.4.4 *Multi-objective Optimization*

6.4.4.1 Decision Variables

As decision variables, in this case study, were selected feed, f, and cutting speed, v. Both were defined for the ranges between the minimum and maximum experimental levels, *i.e.*, in the range where the adjusted functions are valid:

$$0.05 \leq f \leq 0.15; \tag{6.12a}$$

$$80 \leq v \leq 220. \tag{6.12b}$$

6.4.4.2 Objective Functions

As objective functions were selected the surface roughness, R_a, and the tool wear, V_C. They depend upon the cutting parameters, f and v, and cutting time, t, as:

$$R_a = \varphi_1(v, f, t); \tag{6.13}$$

$$V_C = \varphi_2(v, f, t); \tag{6.14}$$

where φ_1 and φ_2 are the neural-network-based models obtained for R_a and V_C, respectively.

On the other hand, cutting time can be computed as:

$$t = \frac{L}{n\,f}; \tag{6.15}$$

where L is the cutting length, and n the rotation speed of the spindle, which can be determined as:

$$n = \frac{1000v}{\pi D}; \tag{6.16}$$

where D is the diameter of the machined surface.

6.4.4.3 Constraints

The single considered constraint is the cutting power, P_C, which must be less than the allowable power given by the motor, P_M:

$$P_C = \frac{F_C v}{6\times 10^4} \leq P_M; \tag{6.17}$$

where the cutting force, F_C, is computed as:

$$F_C = K_S f\, a_p; \tag{6.18}$$

and the specific cutting force, K_S:

$$K_S = \varphi_3(v, f, t); \tag{6.19}$$

where φ_3 is the corresponding neural model.

6.4.4.4 Multi-objective Genetic Algorithm

In order to carry out the optimization process, a multi-objective GA was implemented. The block diagram of the GA is shown in Figure 6.20.

The first step is to create an initial population, of size N. For each individual, it is assigned a value for each decision variable. These values are randomly selected from the respective valid ranges, *i.e.*:

$$x_1 = \mathrm{rnd}(v): 80 \le v \le 220; \tag{6.20a}$$

$$x_2 = \mathrm{rnd}(f): 0.05 \le f \le 0.15. \tag{6.20b}$$

The two values of each individual are encoded to form a code string that represents itself. This code string, the so-called 'chromosome', is composed of binary elements (0 or 1), and has 64 characters (32 for each decision variable).

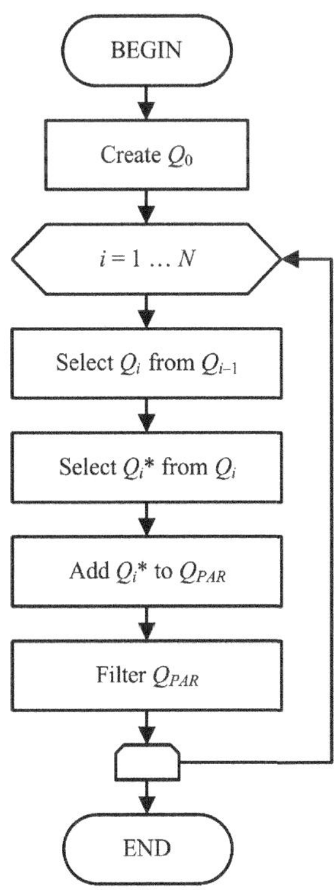

Figure 6.20 Block diagram of the GA

For each individual in the population, the objective functions (surface roughness and tool wear) are evaluated, using the corresponding values of cutting parameters (decision variables).

The constraint is also evaluated. In order to facilitate handling, this constraint is scaled resulting in the form:

$$g_1 = \frac{P_C}{P_M} - 1 \leq 0. \tag{6.21}$$

From this point, a loop is carried out. In each step of this loop (so-called epochs), a new population is created, from the previously existing one, by applying selection, crossover and mutation.

In this approach, selection was carried out by tournament. To create each individual for the new population, two pairs of candidates for parents must be randomly selected. In each pair, both candidates are compared, taking into account the following rules:

- A feasible individual is always better than an unfeasible one.
- In a pair of feasible individuals, one of them is better if it dominates the other one. If no individual dominates the other, both are equally good.
- In a pair of unfeasible individuals, the best is the one which has the smaller unfeasibility index.

The crossover operator combines the code strings of the two successful candidates (one for each pair). In the proposed approach, a two-point crossover is implemented, because in this way it is less disruptive than a multipoint crossover, and it helps the preservation of diversity better than the single-point one.

Finally, by mutation it is possible to obtain some random changes in the code string of the new individuals. This is a technique that helps to introduce new features in the population. Of course, there is no guarantee that these new features could be advantageous; therefore, the mutation likelihood should be kept very low as the high value will destroy good individuals, and degenerates the GA into a random search method. In the proposed GA, a mutation likelihood of 10^{-4} was selected.

The maintenance of an elitist population (so-called Paretian population) is a common technique to preserve the fittest individuals. At the end of each epoch, the non-dominated individuals (Paretian solutions) are selected from the current population and added to the elitist one.

After the addition of new individuals, the elitist populations should be filtered in order to eliminate dominated and staked individuals.

6.4.4.5 Results and Discussion

The optimization was carried out as an *a posteriori* approach, *i.e.*, executing the optimization process in order to obtain a set of non-dominated solutions, and then, making the decision about which of these solution is the most convenient for the specific considered condition.

In Table 6.7, the set of non-dominated solutions, for this case study, are shown. As can be seen, the constraint (cutting power), is notably under the established limit.

These non-dominated solutions are arranged in a Pareto front, in Figure 6.21. This graphical representation helps to make a decision. For example, if there is need to establish a process where a minimum tool wear is achieved and surface roughness is not an important factor, then, point A is the most convenient. On the contrary, if it is desired to obtain the best surface quality, without taking into account the tool wear, then, point B must be selected. In an intermediate

Table 6.7 Outcomes of the optimization process

f (mm/rev)	v (m/min)	t (min)	V_C (mm)	R_a (μm)	Fc (N)	Pc (kW)
0.11	164	10.9	0.04	1.05	70.8	0.19
0.14	142	10.0	0.04	1.03	76.9	0.18
0.15	128	10.8	0.05	1.00	78.5	0.17
0.12	197	9.0	0.10	0.98	73.0	0.24
0.11	181	10.6	0.12	0.96	70.1	0.21
0.11	194	9.7	0.14	0.94	71.7	0.23
0.11	195	9.9	0.16	0.92	71.3	0.23
0.10	194	10.3	0.19	0.90	70.7	0.23
0.10	201	10.0	0.21	0.88	71.4	0.24
0.10	208	9.6	0.22	0.87	72.3	0.25
0.10	214	9.5	0.26	0.85	73.1	0.26
0.10	216	9.5	0.27	0.84	73.1	0.26
0.09	214	10.1	0.32	0.82	72.0	0.26
0.09	216	10.2	0.35	0.81	71.9	0.26

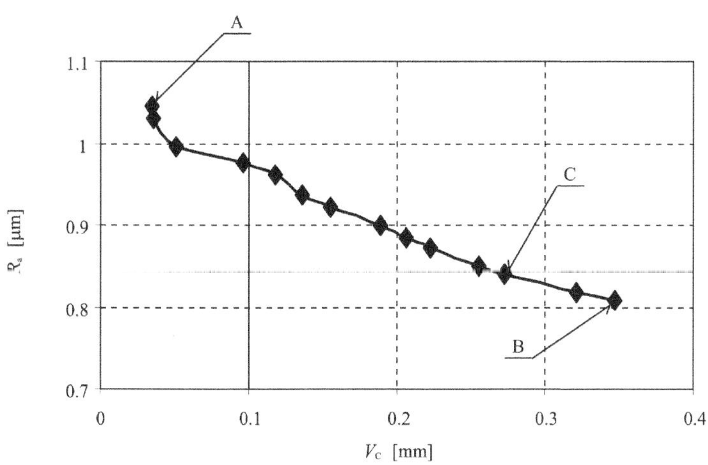

Figure 6.21 Pareto front graph

case, where a value of surface roughness was pre-established (for example, $R_a = 0.84$), the nearest point, in the curve (point C), will be chosen.

6.5 Future Trends

In the near future, an increase in the application of intelligent techniques to hard-machining modelling and optimization can be foreseen. Neural networks and fuzzy logic will be broadly used, because their capability for matching complex relationships. In the same sense, stochastic optimization approaches will be more widely applied for this purpose. This rise will be caused mainly by the continuous increase in the computation power of the computers.

However, all of these tools are currently too green. More solid mathematical foundations are required for this target. Rigorous procedures for setting up and training these approaches and statistical tools for analysing their outcomes must be developed, in order to enhance the effectiveness and reliability of their application.

References

[1] Wang X, Wang W, Huang Y, Nguyen N, Krishnakumar K (2008) Design of neural network-based estimator for tool wear modeling in hard turning. J Intell Manuf 19:383–396

[2] Quiza R, Figueira L, Davim JP (2008) Comparing statistical models and artificial neural networks on predicting the tool wear in hard machining D2 AISI steel. Int J Adv Manuf Technol 37:641–648

[3] Umbrello D, Ambrogio G, Filice L, Shivpuri R (2008) A hybrid finite element method-artificial neural network approach for predicting residual stresses and the optimal cutting conditions during hard turning of AISI 52100 bearing steel. Mater Des 29:873–883

[4] Al-Ahmari AMA (2007) Predictive machinability models for a selected hard material in turning operations. J Mater Process Technol 190:305–311

[5] Singh D, Rao PV (2007) A surface roughness prediction model for hard turning process. Int J Adv Manuf Technol 32:1115–1124

[6] Davim JP, Figueira L (2007) Comparative evaluation of conventional and wiper ceramic tools on cutting forces, surface roughness, and tool wear in hard turning AISI D2 steel. Proc Inst Mech Eng B J Eng Manuf 221:625–633

[7] Basak S, Dixit US, Davim JP (2007) Application of radial basis function neural networks in optimization of hard turning of AISI D2 cold-worked tool steel with a ceramic tool. Proc Inst Mech Eng B J Eng Manuf 221:987–998

[8] Lajis MA, Karim ANM, Amin AKMN, Hafiz AMK, Turnad LG (2008) Prediction of tool life in end milling of hardened steel AISI D2. Eur J Sci Res 21:592–602

[9] Bissey-Breton S, Poulachon G, Lapujoulade F (2006) Integration of tool geometry in prediction of cutting forces during milling of hard materials. Proc Inst Mech Eng B J Eng Manuf 220:579–587

[10] Mamalis A, Kundrák J, Markopoulos A, Manolakos D (2008) On the finite element modelling of high speed hard turning. Int J Adv Manuf Technol 38:441–446

[11] Özel T, Karpat Y, Figueira L, Davim JP (2007) Modelling of surface finish and tool flank wear in turning of AISI D2 steel with ceramic wiper inserts. J Mater Process Technol 189:192–198
[12] Arsecularatne JA, Zhang LC, Montross C, Mathew P (2006) On machining of hardened AISI D2 steel with PCBN tools. J Mater Process Technol 171:244–252
[13] Özel T, Hsu T-K, Zeren E (2005) Effects of cutting edge geometry, workpiece hardness, feed rate and cutting speed on surface roughness and forces in finish turning of hardened AISI H13 steel. Int J Adv Manuf Technol 25:262–269
[14] Özel T, Karpat Y (2005) Predictive modeling of surface roughness and tool wear in hard turning using regression and neural networks. Int J Mach Tools Manuf 45:467–479
[15] Poulachon G, Moisan A, Jawahir IS (2001) Tool-wear mechanisms in hard turning with polycrystalline cubic boron nitride tools. Wear 250:576–586
[16] Noorul Haq A, Tamizharasan T (2006) Investigation of the effects of cooling in hard turning operations. Int J Adv Manuf Technol 30:808–816
[17] Aslan E, Camuscu N, Birgören B (2007) Design optimization of cutting parameters when turning hardened AISI 4140 steel (63 HRC) with Al2O3 + TiCN mixed ceramic tool. Mater Des 28:1618–1622
[18] Sharma VS, Dhiman S, Sehgal R, Sharma SK (2008) Estimation of cutting forces and surface roughness for hard turning using neural networks. J Intell Manuf 19:473–483
[19] Quiza R, Davim JP (2009) Computational modeling of machining systems. In: Özel T, Davim JP (eds) Intelligent machining: modeling and optimization of the machining processes and systems. London: ISTE, pp 173–213
[20] Scheffer C, Kratz H, Heyns PS, Klocke F (2003) Development of a tool wear-monitoring system for hard turning. Int J Mach Tools Manuf 43:973–985
[21] Kothamasu R, Huang S, Verdiun H (2005) Comparison of computational intelligence and statistical methods in condition monitoring for hard turning. Int J Prod Res 43:597–610
[22] Horng J-T, Chiang K-T (2008) A grey and fuzzy algorithms integrated approach to the optimization of turning Hadfield steel with Al2O3/TiC mixed ceramic tool. J Mater Process Technol 207:89- 97
[23] Huang LH, Chen JC (2004) A fuzzy-nets-based in-process surface roughness prediction system in turning operations. Int J Knowl Based Intell Eng Syst 8:37–44
[24] Sun J, Rahman M, Wong YS, Hong GS (2004) Multiclassification of tool wear with support vector machine by manufacturing loss consideration. Int J Mach Tools Manuf 44:1179–1187
[25] Tang Y (2006) Optimization strategy in end milling process for high speed machining of hardened die/mold steel. J Univ Sci Technol Beijing Miner Metall Mater 13:240–243
[26] Stoic A, Kopac J, Cukor G (2005) Testing of machinability of mould steel 40CrMnMo7 using genetic algorithm. J Mater Process Technol 164–165:1624–1630
[27] Xueping Z, Erwei G, Liu CR (2009) Optimization of process parameter of residual stresses for hard turned surfaces. J Mater Process Technol 209:4286–4291
[28] Horng J-T, Liu N-M, Chiang K-T (2008) Investigating the machinability evaluation of Hadfield steel in the hard turning with Al2O3/TiC mixed ceramic tool based on the response surface methodology. J Mater Process Technol 208:532–541
[29] Özel T, Karpat Y (2007) Multi-objective optimization for turning processes using neural network modeling and dynamic-neighborhood particle swarm optimization. Int J Adv Manuf Technol 35(3–4):234–247
[30] Bouacha K, Yallese MA, Mabrouki T, Rigal J-F (2010) Statistical analysis of surface roughness and cutting forces using response surface methodology in hard turning of AISI 52100 bearing steel with CBN tool. Int J Refract Met Hard Mater 28(3):349-361
[31] Paiva AP, Ferreira JR, Balestrassi PP (2007) A multivariate hybrid approach applied to AISI 52100 hardened steel turning optimization. J Mater Process Technol 189:26–35
[32] Chen JB, Huang Y-K, Chen M-S (2005) Feedrate optimization and tool profile modification for the high-efficiency ball-end milling process. Int J Mach Tools Manuf 45:1070–1076

[33] Ozcelik B, Oktem H, Kurtaran H (2005) Optimum surface roughness in end milling Inconel 718 by coupling neural network model and genetic algorithm. Int J Adv Manuf Technol 27:234–241

[34] Davim JP, Figueira L (2007) Machinability evaluation in hard turning of cold work tool steel (D2) with ceramic tools using statistical techniques. Mater Des 28:1186–1191

Index

M

N

O

P

R

S

T

GPSR Compliance

The European Union's (EU) General Product Safety Regulation (GPSR) is a set of rules that requires consumer products to be safe and our obligations to ensure this.

If you have any concerns about our products, you can contact us on ProductSafety@springernature.com

In case Publisher is established outside the EU, the EU authorized representative is:

Springer Nature Customer Service Center GmbH
Europaplatz 3
69115 Heidelberg, Germany

Batch number: 10614906

Printed by Printforce, the Netherlands